THE NATIONAL ACADEMIES

National Academy of Sciences
National Academy of Engineering
Institute of Medicine
National Research Council

The National Academy of Sciences is a private, nonprofit, self-perpetuating society of distinguished scholars engaged in scientific and engineering research, dedicated to the furtherance of science and technology and to their use for the general welfare. Upon the authority of the charter granted to it by the Congress in 1863, the Academy has a mandate that requires it to advise the federal government on scientific and technical matters. Dr. Bruce Alberts is president of the National Academy of Sciences.

The National Academy of Engineering was established in 1964, under the charter of the National Academy of Sciences, as a parallel organization of outstanding engineers. It is autonomous in its administration and in the selection of its members, sharing with the National Academy of Sciences the responsibility for advising the federal government. The National Academy of Engineering also sponsors engineering programs aimed at meeting national needs, encourages education and research, and recognizes the superior achievements of engineers. Dr. Wm. A. Wulf is president of the National Academy of Engineering.

The Institute of Medicine was established in 1970 by the National Academy of Sciences to secure the services of eminent members of appropriate professions in the examination of policy matters pertaining to the health of the public. The Institute acts under the responsibility given to the National Academy of Sciences by its congressional charter to be an adviser to the federal government and, upon its own initiative, to identify issues of medical care, research, and education. Dr. Kenneth I. Shine is president of the Institute of Medicine.

The National Research Council was organized by the National Academy of Sciences in 1916 to associate the broad community of science and technology with the Academy's purposes of furthering knowledge and advising the federal government. Functioning in accordance with general policies determined by the Academy, the Council has become the principal operating agency of both the National Academy of Sciences and the National Academy of Engineering in providing services to the government, the public, and the scientific and engineering communities. The Council is administered jointly by both Academies and the Institute of Medicine. Dr. Bruce Alberts and Dr. Wm. A. Wulf are chairman and vice-chairman, respectively, of the National Research Council.

COMMITTEE ON ECOSYSTEM EFFECTS OF FISHING: PHASE 1—EFFECTS OF BOTTOM TRAWLING ON SEAFLOOR HABITATS

JOHN STEELE, *Chair*, Woods Hole Oceanographic Institution, Massachusetts
DAYTON LEE ALVERSON, Natural Resource Consultants, Seattle, Washington
PETER AUSTER, University of Connecticut, Groton
JEREMY COLLIE, University of Rhode Island, Narragansett
JOSEPH T. DEALTERIS, University of Rhode Island, Kingston
LINDA DEEGAN, Marine Biological Laboratory, Woods Hole, Massachusetts
ELVA ESCOBAR-BRIONES, Universidad Nacional Autonoma de Mexico, Cuidad Universitaria
STEPHEN J. HALL, Australian Institute of Marine Science, Townsville, Queensland
GORDON H. KRUSE, University of Alaska, Fairbanks
CAROLINE POMEROY, University of California, Santa Cruz
KATHRYN M. SCANLON, U.S. Geological Survey, Woods Hole, Massachusetts
PRISCILLA WEEKS, University of Houston, Clear Lake, Texas

Staff

SUSAN ROBERTS, Study Director
JODI BACHIM, Senior Project Assistant
ABBY SCHNEIDER, NRC Fellow

Preface

The Ocean Studies Board has provided advice to Congress and the National Marine Fisheries Service on a broad range of topics relevant to marine fisheries. *Sustaining Marine Fisheries* dealt broadly with the ecological issues (National Research Council, 1999) while other reports have focused on the science of fisheries management: *Review of Northeast Fishery Stock Assessments* (National Research Council, 1998), *Improving Fish Stock Assessments* (National Research Council, 1999), and *Improving the Collection, Management and Use of Marine Fisheries Data* (National Research Council, 2001). Several reports have reviewed management methods including *Marine Protected Areas: Tools for Sustaining Ocean Ecosystems* (National Research Council, 2001) and *Sharing the Fish: Toward a National Policy on Individual Fishing Quotas* (National Research Council, 2000). An emergent theme in nearly all these reports is the desire to reconcile conflicting demands for protecting marine environments, sustaining fishery yields, and responding to the social and economic interests of the fishery and fishing communities. Such reconciliation must integrate scientific and technical information with the diversity of human priorities. This plurality was a central feature of our workshop meetings in Boston, Galveston, and Anchorage with representatives of the fishing industry, environmental groups and members of state and federal agencies.

Our report has benefited significantly from the presentations and discussions at these meetings. The committee would first like to acknowledge the efforts of those who gave presentations: Frank Almeida, Bill Amaru, Pam Baker, Mike Barnette, Francine Bennis, Eloise Brown, Ralph Brown, Arnie Carr, Anthony Chatwin, Jim Churchill, Felicia Coleman, Cathy Coon, Chris Dorsett, Ben Enticknap, Benny Gallaway, John Gauvin, Caroline Gibson, David Goethel, Don Gordon, Alonzo Hamilton, David Harrington, Bill Hayes, Jon Heifetz, Tom Hill, Teressa Kandianis, Rick Leard, Gary Loverich, Trevor McCabe, Linda Mercer, Chris Oliver, Mike Payne, Mark Powell, Jeff Rester, Andy Rosenberg, Brian Rothschild, Pete Sheridan, Michael Sissenwine, Nils Stolpe, Les Watling, Dave Witherell, and Chris Zeman. These talks helped set the stage for fruitful discussions in the closed sessions that followed.

This report has been reviewed in draft form by individuals chosen for their diverse perspectives and technical expertise, in accordance with procedures approved by the NRC's Report Review Committee. The purpose of this independent review is to provide candid and critical comments that will assist the institution in making its published report as sound as possible and to ensure that the report meets institutional standards for objectivity, evidence, and responsiveness to the study charge. The review comments and draft manuscript remain confidential to protect the integrity of the deliberative process. We wish to thank the following individuals for their participation in their review of this report:

ARNE CARR, Massachusetts Department of Fisheries, Pocasset
FELICIA COLEMAN, Florida State University, Tallahassee
CAROLINE GIBSON, Pacific Marine Conservation Council, Friday Harbor, Washington

DANIEL HUPPERT, University of Washington, Seattle
BRIAN ROTHSCHILD, University of Massachusetts, Dartmouth
MICHAEL SINCLAIR, Bedford Institute of Oceanography, Dartmouth, Nova Scotia, Canada

Although the reviewers listed above have provided many constructive comments and suggestions, they were not asked to endorse the conclusions or recommendations nor did they see the final draft of the report before its release. The review of this report was overseen by **NANCY MARCUS**, Florida State University, Tallahassee, appointed by the Divison on Earth and Life Studies, who was responsible for making certain that an independent examination of this report was carried out in accordance with institutional procedures and that all review comments were carefully considered. Responsibility for the final content of this report

rests entirely with the authoring committee and the institution.

The committee is also grateful to Andrea Dunn for compiling information about the regional fisheries that comprises Table 4.2. Jim Lester also provided important material for this report. The committee wishes to thank the staff at the Ocean Studies Board for their efforts in support of this study: Jodi Bachim, Senior Project Assistant and Abby Schneider, National Research Council Fellow. The committee is especially grateful to the Study Director Dr. Susan Roberts who not only had the task of integrating the individual contributions from committee members into a coherent report, but also guided the report through the extensive process of review and revision.

John Steele
Chair

Contents

Executive Summary

Fishing has a variety of effects on marine habitats and ecosystems, depending on the spatial extent of fishing, the level of fishing effort, and the type of gear. Expansion of U.S. domestic fisheries after passage of the Magnuson-Stevens Fishery Conservation and Management Act of 1976 fueled advances in gear and navigation technology that greatly increased the geographic extent of these effects. However, declining fish stocks have reduced fishing activities in some areas over the past decade. After passage of the Sustainable Fisheries Act in 1996, which required that fishery management plans address the effects of fishing on habitat, attention focused on how fishing affects the seafloor. The primary fisheries involved in the controversy are trawl and dredge fisheries, which tow gear over seafloor habitats and communities. A complete consideration of the effects of fishing on ecosystems would require evaluation not only of trawl and dredge gear, but also of stationary gear (traps, pots, longlines, gillnets) and other kinds of towed gear (pelagic trawls) on target and nontarget species.

As a first step in evaluating the ecosystem effects of fishing, the National Marine Fisheries Service (NMFS) asked the Ocean Studies Board of the National Academy of Sciences to study the effects of bottom trawling and dredging on seafloor habitats. Specifically, NMFS asked the committee to undertake the following tasks: 1) summarize and evaluate existing knowledge on the effects of bottom trawling on the structure of seafloor habitats and on the abundance, productivity, and diversity of bottom-dwelling species in relation to gear type and trawling method, frequency of trawling, bottom type, species, and other important characteristics;

2) summarize and evaluate knowledge about changes in seafloor habitats associated with trawling and with the cessation of trawling; 3) summarize and evaluate research on the indirect effects of bottom trawling on non-seafloor species; 4) recommend how existing information could be used more effectively in managing trawl fisheries; and 5) recommend research to improve understanding of the effects of bottom trawling on seafloor habitats.

During the study, the committee held public meetings in several regions with participation by fishery scientists and managers, the fishing industry, and environmental groups. Discussions often centered on concerns that habitat protection initiatives would become avenues for the reallocation of resources among stakeholders, including various sectors of the fishing industry, recreational fishing groups, and conservation organizations. Resolution of these allocation considerations to meet ecological and socioeconomic goals often has been contentious.

The policy context for addressing the effects of fishing on habitat is found in the essential fish habitat (EFH) provisions specified by the 1996 Sustainable Fisheries Act amending the Magnuson-Stevens Fishery Conservation and Management Act. The amended act requires regional fishery management councils to describe and identify EFH for each fish stock managed under a fishery management plan, to minimize to the extent practicable adverse effects on such habitat caused by fishing, and to identify other actions to encourage habitat conservation and enhancement. Instead of amending individual fishery management plans, most regional councils developed a single, overarching EFH amendment. The Secretary of Commerce approved most of the revised plans, but some environmental groups have mounted legal challenges regarding the adequacy of some EFH amendments. A major complaint was that the regional councils did not sufficiently address the effects of fishing gear on benthic habitats.

Gaps in existing knowledge of the distribution of benthic life stages of fishes and other species and of the physical and biological characteristics of the seafloor made it difficult for the regional councils to develop criteria for designating EFH. Similarly, the councils struggled with the requirement to assess the effects of bottom trawling and dredging because of insufficient data on the spatial scale and extent of bottom fishing. The councils also lacked guidelines for generalizing the results of research on specific gears and habitats. These problems relate to the committee's task to recommend ways of using existing information in the management of the habitat effects of trawl and dredge fisheries.

A complete assessment of the ecosystem effects of trawling and dredging requires three types of information:

1) gear-specific effects on different habitat types (obtained experimentally);
2) frequency and geographic distribution of bottom tows (trawl and dredge fishing effort data); and
3) physical and biological characteristics of seafloor habitats in the fishing grounds (seafloor mapping).

This report summarizes current data in these three areas and describes how the low spatial resolution and availability of the fishing effort and habitat mapping data restrict a full evaluation of the ecosystem effects of trawling and dredging.

Under the first category of information, many experimental studies have documented the acute, gear-specific effects of trawling and dredging on various types of habitat. The results confirm predictions based on the ecological principle that stable communities of low mobility, long-lived species will be more vulnerable to acute and chronic physical disturbance than will short-lived species in changeable environments. Trawling and dredging can reduce habitat complexity by removing or damaging the biological and physical structures of the seafloor. The extent of the initial effects and the rate of recovery depend on the habitat stability. The more stable biogenic (i.e., of biological

origin), gravel, and mud habitats experience the greatest changes and have the slowest recovery rates. In contrast, less consolidated coarse sediments in areas of high natural disturbance show fewer initial effects. Because those habitats tend to be populated by opportunistic species that recolonize more rapidly, recovery is faster as well. Significant alterations to habitat can cause changes in the associated biological communities, potentially altering the composition and productivity of fish communities that depend on seafloor habitats for food and refuge.

The second category of information, the geographic distribution and frequency of trawling and dredging, suffers from limitations in the spatial resolution of the data and in regional variation in reporting methods. For example, trawling effort data are averaged over reporting areas that range $25–2420$ km^2, depending on the region. Although the data are imperfect, a few generalizations emerged from the analysis presented in this report. Based on estimates of the spatial extent and intensity of trawl and dredge fishing effort, bottom trawling takes place over large areas of the continental shelf and slope. The level of effort varies greatly among regions. The highest intensity of effort, based on rough estimates of the number of times a reporting area is swept (Table 4.1), occurs in the fishing grounds of the Gulf of Mexico and New England regions. In contrast, bottom trawling in the mid-Atlantic, Pacific, and North Pacific regions is relatively light, with less than one tow per year in many reporting areas. Even in heavily trawled regions, effort is not evenly distributed. As a consequence, some areas may be trawled several times per year while other areas may be trawled infrequently if at all. Throughout the 1990s and into 2001 there were significant reductions in the intensity and spatial extent of bottom trawling. Those reductions reflect effort reductions, area closures, and gear restrictions instituted by managers in response to problems with declining fish stocks, bycatch, or interactions with endangered species.

The spatial distribution of different habitat types in trawled (or dredged) areas is the third category of information that must be integrated with the other two to assess the effects of trawling and dredging on ecosystems. Experimental studies on specific gear types in a few well-defined habitats provide small-scale estimates of ecological disturbance, but for most areas only coarse maps are available on habitat distribution.

The mismatch in the spatial scales of experimental results, habitat maps, and trawl effort reporting data makes it difficult to assess the ecosystem-level effects of trawling and dredging. Although fisheries managers collect data continuously, limitations in resources and time require them to assess effects in the absence of complete information. In this context, comparative risk assessment provides a promising approach to evaluate the effects of bottom trawling and dredging. This method brings together the various stakeholders to identify risks to seafloor habitats and to rank management actions within the context of current statutes. Because risk assessment requires full use of all available information on seafloor habitats, fishing methods, and effort distribution there is an immediate need to integrate the available data in a readily available format.

RECOMMENDATIONS

Although there are still habitats, gears, and geographic regions that have not been adequately studied and characterized, there is an extensive literature on the effects of fishing on the seafloor. It is both possible and necessary to use this existing information to more effectively manage the effects of fishing on habitat. The following recommendations fall into three categories: 1) interpretation and use of existing data; 2) integration of management options; and 3) policy issues raised by existing legislation. These recommendations are intended to build upon the strengths of existing approaches to management rather than completely transform them.

Interpretation and Use of Existing Data

Recommendation

Fishery managers should evaluate the effects of trawling based on known responses of specific habitat types and species to disturbance by different fishing gears and levels of fishing effort, even when region-specific studies are not available. The lack of area-specific studies on the effect of trawling and dredging gear is insufficient justification to postpone management of fishing effects on seafloor habitat. The direct responses of benthic communities to trawling and dredging are consistent with ecological predictions based on disturbance theory. Predictions from common trends observed in other areas provide useful first-order approximations of fishing effects for use in habitat management. As more site-specific information becomes available on the fine scale distribution of

fishing effort and habitat distribution, those estimates should be revised.

Recommendation

NMFS and its partner agencies should integrate existing data on seabed characteristics, fishing effort, and catch to provide geographic databases for major fishing grounds. Management decisions about how fishing affects habitat can be improved by the simultaneous and consistent presentation of all available data on the characteristics of the seabed and fishing effort. There are data that describe seabed types and habitats and the location and intensity of fishing for much of the U.S. continental shelf. Available data sets collected by different agencies currently exist in different formats, at variable levels of resolution, in separate archives. Their integration into a single geographic information system will help managers evaluate regional needs for habitat conservation.

Integration of Management Options

Recommendation

Management of the effects of trawling and dredging should be tailored to the specific requirements of the habitat and the fishery through a balanced combination of the following management tools.

- *Fishing effort reductions.* Effort reduction is the cornerstone of managing the effects of fishing, including, but not limited to, effects on habitat. Both of the other management tools (gear restrictions or modifications and closed areas) also can require effort reduction to achieve maximum benefit. The success of fishing effort reduction measures will depend on the resilience and recovery potential of the habitat.
- *Modifications of gear design or gear type.* Gear restrictions or modifications that minimize bottom contact can reduce habitat disturbance. Shifts to different gear types or operational modes can be considered, but the social, economic, and ecological consequences of gear reallocation should be recognized and addressed.
- *Establishment of areas closed to fishing.* Closed areas are necessary to protect a range of vulnerable, representative habitats. Closures are particularly useful for protecting biogenic habitats (corals, bryozoans, hydroids, sponges, seagrass beds) that are disturbed by even minimal fishing

effort. Because area closures could displace effort to open fishing grounds, effort reductions could be necessary in some cases to reduce habitat effects.

The optimal combination of these management approaches will depend on the characteristics of the ecosystem and the fishery—habitat type, resident sea-floor species, frequency and distribution of fishing effort, gear type and usage, and the socioeconomics of the fishery. Each characteristic should be considered during development of management plans for mitigating the impacts of fishing on the seafloor.

Recommendation

The regional fishery management councils should use comparative risk assessment to identify and evaluate risks to seafloor habitat and to prioritize management actions within the context of current statutes and regulations. Risk assessment, in general, is a scientifically informed way of clarifying public debates over environmental policy by making explicit the environmental consequences of particular policy choices. Comparative risk assessment provides the following advantages for the task of benthic habitat protection:

- It can be used even in the absence of scientific certainty because it relies on a combination of available data, scientific inference, and public values.
- It provides simultaneous analysis of a wide range of risks to benthic habitats. Mobile bottom gear is only one of many factors that contribute to the degradation of benthic habitats. Other factors might include pollution, drilling and natural disturbance.
- It enables stakeholder involvement in the decision-making process.

Policy Issues Raised by Existing Legislation

Recommendation

Guidelines for designating EFH and habitat areas of particular concern (HAPC) should be established based on standardized ecological criteria. The underlying aim of the EFH concept is valuable, and it appropriately emphasizes the need to place management of exploited fishes within the context of managing the total ecosystem. The current designation of EFH, however, does not require the use of consistent criteria for the assignment of habitat to each life stage of the species covered by fishery management plans. Instead, the regional councils develop the criteria, often based on data availability. Current EFH designations typically are too extensive to form a practical basis for managing fisheries. Although this approach could help mitigate some threats to habitat, it provides little guidance for evaluating the consequences of trawling and dredging. EFH designations should be based on a clear understanding of the population biology and spatial distribution of each species.

An HAPC constitutes a subset of EFH based on the ecological value of the area, its susceptibility to perturbation, and whether it is rare or stressed (National Marine Fisheries Service, 1997). HAPCs require the strongest safeguards to ensure habitat protection because their value in the life cycle of exploited fish populations has been documented. Nevertheless, no such protection is afforded in the current policy structure. HAPC should be clearly and narrowly defined, specific guidelines should be set for determining permissible activities, and a schedule for reviewing the effectiveness of the designation should be developed.

Recommendation

A national habitat classification system should be developed to support EFH and HAPC designations. Efforts to inventory and construct regional or national habitat maps require a classification system with common designations. Such a system would facilitate tracking of changes over time and would provide a basis for determining functional links between seafloor ecosystems and fisheries production. A classification system would assist in ranking different habitats according to the resilience of their biological communities and associated fisheries; estimating habitat vulnerability; and managing habitat impacts based on the generalized results of research conducted in other geographic areas.

FUTURE RESEARCH

Many gaps were identified in current understanding of the consequences of fishing on the seafloor. The following recommendations are intended to direct research toward filling these gaps. They have been organized into three primary areas of research—gear effects and modification, habitat evaluation, and management—with some overlap between categories.

Gear Effects and Modification

Fishermen's knowledge and experience should be used to study gear impacts and to develop new gear technology. Their active engagement in research will help ensure that mitigation strategies are practical, enforceable, and acceptable to the fishing community. Further research on gear effects will be required to develop a predictive capability to link gear type and effort to bottom disturbance, fish production, and recovery times in particular habitats. New research is needed in the following areas:

- identification of the forces that injure and dislodge a range of benthic organisms;
- development of fishing gear that is less damaging to habitat and that helps meet other conservation goals, such as bycatch reduction and maintenance of biological communities; and
- determination of the relationship between fish production and bottom disturbance, especially for areas that continue to support fish despite chronic disturbance by fishing gear.

Habitat Evaluation

Most previous research has addressed habitat disturbance on a small spatial scales. The focus has been on short-term, acute disturbance, and on animal communities rather than ecosystem processes (productivity, nutrient regeneration). Closed areas should be used as control sites to study the chronic effects of seabed disturbance by trawl or dredge gear. Future research should examine:

- cumulative effects of trawling on sites that have been trawled repeatedly;
- repeated disturbances by fishing gears to determine the dose-response relationship as a function of gear, recovery time, and habitat type;
- recovery dynamics, with estimates of large-scale effects at current fishing intensities;
- acute and chronic effects of trawling in deeper water (>100 m); and
- recovery rates in stable and structurally complex habitats for which the return time will be measured in years to decades.

Evaluation of the indirect effects of bottom trawling and dredging will require experimentation, modeling,

and comparison of different habitat types to analyze trends in benthic production and community structure relative to trends in fisheries production. This evaluation should include:

- effects of habitat fragmentation on biological communities and the productivity of exploited fish stocks;
- rates and magnitude of sediment resuspension, nutrient regeneration, and responses of the plankton community in relation to gear-induced disturbance; and
- long-term trend data on benthic production versus fisheries production.

Management

Productive interactions among stakeholders and policymakers should be enhanced through increased participation in research on the effects of fishing on the seafloor and development of alternative gears and practices. Interactions can be facilitated through user group funding of research and by collaborative research projects that involve scientists and fishermen.

Development of better quantitative data for risk analysis will require research on the habitats and population dynamics of nontarget species, specifically:

- adequate baselines for particular habitats and regions, to document the effects of various fishery practices;
- testable hypotheses about how communities in different habitat will respond to fishing;
- quantitative models to predict fishing effects in areas that have not been studied; and
- mortality estimates for nontarget species.

NMFS should establish protocols for studying existing trawl and dredge area closures to evaluate ecological, social, and economic effects of habitat management strategies. This will aid assessment of management alternatives in other locations. Aggregation and analysis of existing information on habitats, fishing effort, and efficacy of various management measures will help the regional fishery management councils meet their mandate to protect EFH. Research that will facilitate management decisions include:

- analysis of community structure and life history parameters to validate the use of frequency dependent distribution approaches for designating EFH and HAPC; and

• collection and analysis of data on the social and economic characteristics of trawl, dredge, and nonmobile gear fisheries to assess the tradeoffs among various management alternatives.

CONCLUSION

Integration of available data on the effects of trawling and dredging, fishing effort, and the distribution of seafloor habitats can provide a starting point for practical initial evaluations that will inform management decisions. Management measures should be assessed regularly to provide better information about how various restrictions affect fish stocks and habitats and to determine the socioeconomic effects on the fishing industry and fishing communities.

However, existing data are not sufficient to optimize the spatial and temporal distribution of trawling and dredging to protect habitat and sustain fishery yields. Resolution of the different, and at times conflicting, ecological and socioeconomic goals will require not only a better understanding of the relevant ecosystems and fisheries, but also more effective interaction among stakeholders.

1

Introduction

Over the last 25 years there has been increasing interest in the effects of fishing—not only for target species but also for nontarget species. The environmental consequences—how fishing affects habitat—have come into the discussion during the past decade. Numerous studies indicate that habitat complexity improves the survivorship of many fish species. Benthic organisms (plants, corals, and sponges) and sediment forms (mud burrows and gravel) add structure to the seafloor and increase habitat complexity. Seafloor structures serve as nurseries for juvenile fish and provide refuge and food for adults. Even small structures, such as cobbles and clam shells, can form important habitat. Areas of the seafloor that lack these structures do not support the variety of fish populations observed in more complex regions (Collie et al., 1997; Kaiser et al., 1999).

Fishing affects marine habitats and ecosystems in many ways that depend on the type of gear used and on the spatial and temporal extent of fishing (Auster and Langton, 1999). Mobile fishing gear, such as bottom trawls and dredges, can be configured to drag across the seafloor to catch demersal fish and shellfish and some semipelagic fish. But this method of fishing disturbs the structure of the seafloor, affecting the three-dimensional character and availability of fish habitat and changing the composition of biologic communities in the area. The indirect effects of trawling and dredging include disruption of the food web, alteration in the rate of decomposition of organic matter, and recycling of nutrients through resuspension of bottom sediments.

It is difficult to quantify the effects of habitat disruption caused by trawling and dredging because there are so many sources of disturbance: depletion of commercial and recreational fish stocks, pollution, changes in climate, and oceanographic variability. Those factors complicate the evaluation of the general consequences of fishing on marine ecosystems, including the specific effects of habitat degradation. To study the effects of fishing within a background of variability from other sources researchers often compare unfished areas with areas that are experimentally trawled or dredged. The difficulty is in finding analogous untrawled control areas. It also has been difficult to quantify, and therefore to generalize, the relationships among fishing intensity and frequency, fishing methods and gear, seafloor structure, productivity and abundance of economically valuable species, and diversity of other organisms. Finally, some fishermen, fishery managers, and scientists have expressed skepticism about the validity of generalizing data from one region to another as a basis for implementing regulations.

In the 1996 reauthorization of the Magnuson-Stevens Fishery Conservation and Management Act—the Sustainable Fisheries Act (SFA)—Congress increased the regulatory focus on habitat protection through the inclusion of essential fish habitat (EFH) provisions. Regional fishery management councils (FMCs), which manage most of the marine fisheries in the United States, were required by the act to "describe and identify essential fish habitat" for each managed fish stock "to minimize to the extent practicable adverse effects on such habitat caused by fishing." FMCs must evaluate the effects of all fishing practices on seafloor habitat.

The National Oceanic and Atmospheric Administration asked the Ocean Studies Board of the National

---

Box 1.1
Statement of Task

This study will be the first in a series that will evaluate available data related to the physical and biological effects of fishing on marine habitats and ecosystems. This first study will 1) summarize and evaluate existing knowledge on the effects of bottom trawling on the structure of seafloor habitats and the abundance, productivity, and diversity of bottom-dwelling species in relation to gear type and trawling method, frequency of trawling, bottom type, species, and other important characteristics; 2) summarize and evaluate knowledge about changes in seafloor habitats with trawling and cessation of trawling; 3) summarize and evaluate research on the indirect effects of bottom trawling on non-seafloor species; 4) recommend how existing information could be used more effectively in managing trawl fisheries; and 5) recommend research needed to improve understanding of the effects of bottom trawling on seafloor habitats.

Research Council's Division on Earth and Life Sciences to undertake a series of studies to examine the effects of various fishing practices and to make recommendations for action that could reduce or mitigate effects. This first report addresses the specific effects of bottom trawling on seafloor habitats, as described in the statement of task (Box 1.1). In deliberations at its first meeting on the topic, the committee decided to address the effects of both trawls and dredges, because these are the two major types of bottom-tending mobile gear used in U.S. fisheries.

STUDY APPROACH AND REPORT ORGANIZATION

This report summarizes the literature on the effects of bottom trawling on habitats and discusses management tools that can be applied to mitigate them. Inshore and offshore areas are considered, although there is more emphasis on offshore regions because of the greater amount of information generally available for federally managed waters. The Caribbean and West Pacific regions are not discussed because they have no major trawl or dredge fisheries. The regulatory framework provided by provisions of the SFA is introduced

below. Chapter 2 describes different types of mobile bottomtending gear used in trawl and dredge fisheries. Chapter 3 summarizes research findings on the direct and indirect effects of trawling and dredging and it reviews the literature on the postdisturbance recovery of habitat and biota. Chapter 4 describes what is known about the seafloor habitats and about the regional distribution and frequency of trawling and dredging activities. Appendix B presents maps and more detailed descriptions of the distribution and frequency of trawling in U.S. waters. Ecological risk assessment methods are presented in Chapter 5, and management options for reducing the damage caused by trawling and dredging are presented in Chapter 6. In Chapter 7, the committee presents its conclusions and recommends topics for research. Appendix A lists committee and staff members. Appendix C gives a brief explanation of mapping tools.

LEGISLATIVE CONTEXT

EFH provisions of the Sustainable Fisheries Act (1996) gave resource managers a new tool to address degradation and loss of fish habitat. The final rule published by the National Marine Fisheries Service (NMFS) (2002) defines EFH as follows:

> . . . those waters and substrate necessary to fish for spawning, breeding, feeding, or growth to maturity. For the purpose of interpreting the definition of essential fish habitat: "Waters" include aquatic areas and their associated physical, chemical, and biological properties that are used by fish and may include aquatic areas historically used by fish where appropriate; "substrate" includes sediment, hard bottom, structures underlying the waters, and associated biological communities; "necessary" means the habitat required to support a sustainable fishery and the managed species' contribution to a healthy ecosystem; and "spawning, breeding, feeding, or growth to maturity" cover a species full life cycle.

The act requires fishery management plans to describe and identify EFH, minimize to the extent practicable adverse effects on EFH caused by fishing, and identify other actions to encourage habitat conservation and enhancement. EFH must be designated for each life stage of the more than 700 federally managed species. Additionally, NMFS must provide conservation recommendations to all federal or state agencies on actions that adversely affect EFH. Federal agencies must respond within 30 days to recommendations, although the recommendations are nonbinding. Hence, the SFA made habitat conservation a mandate under federal fisheries management.

The regional fishery management councils were charged with identifying EFH for the egg, larva, juvenile, and adult stages of each species they manage, even though in many cases there was little information to assist them. The broad definition of EFH used in the legislation made for a daunting task. Each council took a slightly different approach, developing EFH amendments by species, by multispecies complexes, and by habitat type. The process used by the New England Council illustrates one approach.

The New England Council produced a stand-alone EFH amendment to address the needs of all 18 of its managed species. Amendment development was divided into three distinct phases. In the first and most time-consuming phase, the Council identified and described EFH for managed species. The Council based its descriptions predominantly on NMFS' "EFH Source Documents," a compilation of 31 EFH species reports, consisting mostly of NMFS survey data. The adult and juvenile data were compiled from bottom trawl surveys conducted by NMFS (1963–1997) and the larvae and egg data were based on the Marine Resources Monitoring, Assessment, and Predictions ichthyoplankton survey (1977–1987). EFH often was defined by the Council as the area containing 90–100 percent of each life stage for each managed species. Because time and resources were limited, the Council primarily used these data sets to designate EFH for numerous species.

In the second phase of development, the New England Council evaluated fishing related effects on EFH. This involved extensive literature reviews and drew heavily on "The Effects of Fishing on Fish Habitat" (Auster and Langton, 1999). The Council then identified and reviewed measures to reduce the potential for harm from fishing activities. In the third and final phase, the Council identified a range of actions to mitigate damage from nonfishing related activities, although the Council lacks the authority to regulate those activities. Based on its experience, the Council identified many areas of habitat-related research and information that would aid in the evaluation and improvement of its existing EFH designations.

The approach used by the New England Council defined EFH for each life stage of all managed species and thereby satisfied SFA requirements. However, under the criteria used to identify EFH, maps developed for New England included most of the waters off its coast. Even when EFH for only three species and one life stage (juvenile cod, pollock, and haddock) are plotted on a single map, most of the Gulf of Maine region and Georges Bank is classified as EFH (Figure 1.1).

This dilemma is not unique to New England. Many other councils developed EFH maps that covered most if not all of the areas under their jurisdiction. For example, the Gulf of Mexico Fishery Management Council took a cumulative, species-by-species approach to designating EFH. Once an area had been identified as EFH for one species it was not evaluated. The Gulf of Mexico Council concluded that the entire region under its jurisdiction could be classified as EFH. The South Atlantic Council did not use species abundance as a criterion for designating EFH. Instead, it identified habitat types associated with the managed species and designated EFH based on the presence of those habitat types.

NMFS recognized that the definition of EFH was so broad that the mandate to minimize the effects of fishing would be difficult to implement. So it included the concept of habitat areas of particular concern (HAPC) to provide a focus for conservation efforts. There is no requirement that councils designate HAPC for any species, nor does doing so confer additional protection or restrictions (National Marine Fisheries Service, 1997, 2002). HAPC was not a new concept; it had been developed in response to mandates contained in the 1986 amendments to the Magnuson-Stevens Act. NMFS defined HAPC as "EFH that is judged to be particularly important to the long-term productivity of populations of one or more managed species, or to be particularly vulnerable to degradation" (National Marine Fisheries Service, 1997, 2002). Under the EFH regulations, HAPC can be designated based on one or more of the following criteria:

1) the importance of the ecological function provided by the habitat;
2) the extent to which the habitat is sensitive to human-induced environmental degradation;
3) whether and to what extent development activities are, or will be, stressing the habitat type; and
4) the rarity of the habitat type.

Where there was significant information about the habitat needs of a particular species of fish, some councils defined HAPC. For example, the New England Council designated an area on Georges Bank as HAPC for juvenile Atlantic cod, based on evidence from numerous scientific studies that the gravel–cobble

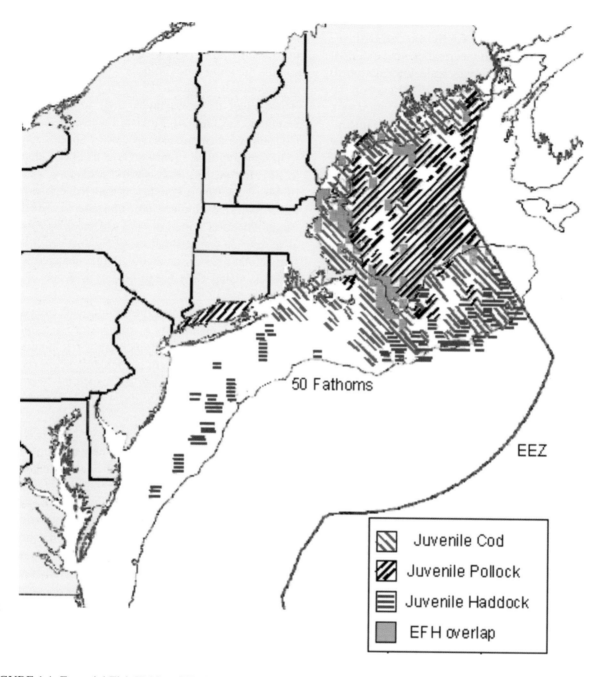

FIGURE 1.1 Essential Fish Habitat (EFH) designations for juvenile cod, pollock, and haddock in New England waters (data from New England Fishery Management Council).

substrate present in this region provides space for the newly settled juvenile cod to find shelter and avoid predation. All councils except the Pacific Fishery Management Council have used HAPC designations to identify areas of special ecological importance. Descriptions of HAPC designations for each region are available from NMFS (www.nmfs.noaa.gov/habitat/ habitatprotection/regionalapproaches.htm).

The concepts of EFH and HAPC provide NMFS and the regional councils with the authority to mitigate habitat degradation from bottom trawling and dredging, among other effects of fishing. Recent controversy and litigation regarding EFH amendments submitted by the regional councils and approved by the Secretary of Commerce has made it timely to examine the extent of the problem and the various management options for mitigating impacts.

IDENTIFYING ESSENTIAL FISH HABITAT

EFH regulations published by NMFS in the interim and final rules describe a four-tiered approach to organizing information for describing and identifying EFH. The tiers are in order of increasing availability of information about the habitat requirements of managed species, as follows:

- *Level 1*: Distribution data are available for some or all of the geographic range of the species. Either systematic presence–absence data or opportunistic observations of the location of various life stages may be used to infer habitat use.
- *Level 2*: Habitat-related densities of the species are available. Geographic information on the density or relative abundance of a species at each life stage may be used to assess habitat value compared with the overall species distribution.
- *Level 3*: Growth, reproduction, or survival rates within habitats are available. The success of the species in a given habitat—based on growth, reproduction, and survival rates—is used as a proxy for productivity.
- *Level 4*: Production rates by habitat are available. Direct assessments of production rates as a function of habitat type, location, quality, and quantity are used to determine the habitat essential for a sustainable fishery and for the species' contribution to a healthy ecosystem.

In most cases, EFH has been designated at Level 2, using frequency-dependent distributions of fishes as a proxy for habitat. That is, essential habitat lies within the region with the highest density of a species. This method is based on sound ecological principles, but often EFH has been designated using the top 90–100 percent of the distributions of many species, based on

---

Box 1.2
Managed Species and Habitat Types

Habitat is that part of the environment on which organisms depend directly or indirectly to carry out life processes. For fish, this includes spawning grounds, nursery areas, feeding areas, and migration routes. Habitat includes the physical environment (structure provided by biogenic animals, plants, and sediments; depth of water), the chemical environment (salinity, dissolved oxygen), and the many organisms (plants, invertebrates) that constitute a food web. Density fronts separate water masses or plumes of turbid, low salinity water produced by large rivers. Kelp beds, seagrass meadows, intertidal marshes, mud, sand or cobble flats, and offshore ledges and banks are distinct areas that serve as habitat for fish and other marine organisms. The physical substrate is often the most noticeable aspect of a habitat and is therefore the basis for many habitat classifications (Langton et al., 1995; Auster, 1998).

---

density or catch per unit effort data as an indicator of abundance. Using more restricted ranges at the top end of the distributions (10–30 percent) would narrow EFH designations to the more preferred habitats, but the optimal range for different species and life stages will require further analysis and definition.

Levels 3 and 4 require significant amounts of information about the relationship between the managed species and the type of habitat (Box 1.2). Fish require a broad diversity of intact habitat functions and processes to survive, grow, and reproduce. Because the physical structure of an area is often the most noticeable aspect of habitat, structure is the basis for most habitat classifications (Allee et al., 2000; Chapter 4) and has been the focus of many studies on the effects of fishing on habitat. However, the biotic component of habitat (food) is equally important to sustaining fish production. Therefore, both the physical and biologic components of seafloor habitat must be included in assessing the effects of trawling and dredging.

2

Characterization of Fishing Gear

Most studies on the effects of fishing gear on the seafloor have been focused on trawl and dredge fisheries. A significant portion of the landings of finfish and shellfish from U.S. coastal waters bottom is made with contact, mobile fishing gear. Bottom trawls catch mostly groundfish and shrimp; dredges are used primarily for scallops and clams. Because different gear types cause different types of seafloor disturbance, it is important to understand the nature of the gear used in the fisheries before considering the effects of those fisheries on habitat. Other gear, such as pelagic trawls and seines, are not included in this chapter because they either have less contact with the seafloor or they represent a minor fraction of the total fishing effort.

Fishing gear that is dragged over the seabed or through the water is called mobile gear. All bottom-contact, mobile fishing gear disturbs the seafloor to some extent. The dragged gear generally includes a bag constructed of synthetic webbing or metal rings and chain links that collect the catch. It is classified as active fishing gear because the animals do not voluntarily enter the gear; they are either swept up from the seabed or netted from the water by the gear. As fishermen strove to fish more efficiently, dragged gear evolved to cover greater areas, or sweeps, of the seafloor. Dragged gear initially was deployed from hand-powered boats, then sailing vessels, and finally from diesel-powered ships, some with engines greater than 1000 horsepower (HP).

Mechanization of fisheries with powerful engines and winches allowed larger gear to be towed faster and handled with less labor. Improved vessel design made it possible for fishermen to travel farther from their home ports and to fish under adverse weather conditions. Improvements in fishing vessel electronic navigation equipment from radio direction finders (LORAN), collision avoidance equipment, radar, and finally satellite-based navigation, such as global positioning systems (GPS), have increased fishing efficiency. The application of sonar technology to fish finding and to mapping the seabed has improved the ability of fishermen to target areas of the seabed that were not previously accessible to dragged gear. Technology also has helped fishermen avoid sites where rough seabed terrain could result in the loss or damage to gear (Box 6.3). Dragged gear has developed from small shellfish rakes towed by hand-paddled canoes to groundfish trawls that are equipped with net mouth openings that in some cases exceed 50 × 100 m.

Trawling off the United States began in the early 1900s, and at about the same time on the Atlantic and Pacific coasts (between 1901 and 1905). By 1971, trawling was considered "a most efficient method of offshore fishing" (Shapiro, 1971). Trawling extended rapidly along the Atlantic coast and into the Gulf of Mexico and became the dominant method of catching fish and shrimp for human consumption. Although trawl fishing also spread along the Pacific coast, the relatively narrow continental shelf and limited West Coast markets led to establishment of a smallboat fishery (generally less than 25 m in length) until the advent in the late 1960s of foreign fisheries.

After World War II, distant water foreign trawl fisheries developed off the New England states, along the Pacific coast, and in the Bering Sea and Gulf of Alaska. Those foreign fisheries were largely phased out by

1990 with the implementation of the Magnuson-Stevens Fisheries Conservation and Management Act (1977), which extended U.S. jurisdiction over fishery resources throughout the newly declared 200 mile exclusive economic zone. Large-scale domestic catcher–trawler and mothership operations arose in the waters off Alaska (Ianelli and Wespestad, 1998) and from the coast of northern California to Washington. In New England, the trawl fisheries also expanded as larger, more efficient vessels were introduced into the fleets. Currently, there are major groundfish trawl fisheries in New England and Alaska. The shrimp trawl fishery dominates in the Gulf of Mexico and in shallow waters off the coasts of the Carolinas and Georgia. Groundfish and shrimp trawl fisheries operate off the coasts of Washington, Oregon, and California.

DREDGE GEAR

Most dredges are rakelike devices that use bags to collect the catch. They typically remove molluscan shellfish from the seabed, but occasionally are used for crustacea, finfish, and echinoderms. Dredges take either epifauna or infauna; the design details of the gear are fauna specific. On soft bottoms, the dredge flattens the microrelief on the seabed (wave-ripples) and resuspends fine sediments. On hard rocky bottoms, the dredge scrapes off epibenthic organisms and disturbs the substrate.

In estuarine waters, dredges are used to collect clams, oysters, conch, and crabs. The oyster dredge consists of a steel frame, 0.5–2.0 m wide, with an eye and "nose" or "tongue" and a blade with teeth. The tow chain or wire and a bag for the catch are attached to the frame. The dredge is towed slowly (<1 m/s) in circles, from vessels that are 7–15 m long. Similar dredges are used to catch blue crabs in the mid-Atlantic region during the winter. Stern-rig dredge boats (averaging 15 m long) drag two dredges in tandem from a single chain warp. The dredges are equipped with long teeth (10 cm) that dig the crabs out of the bottom. The same gear is used in the Chesapeake Bay to catch whelk in summer and mussels in fall.

In the soft clam fishery, which occurs in shallow estuarine waters, the dredge head (manifold and blade) is attached to a conveyor or belt that carries the materials retained on the blade to the working deck of the vessel. These dredge vessels (generally about 15 m long) are restricted to shallow water—less than one-half the length of the escalator. Because soft clams live

in shallow waters (2–6 m), this poses few limitations for the small vessel operators who use this gear.

Offshore, large dredges catch sea scallops, which inhabit a sand–gravel–cobble bottom and live on the surface of the seabed. Because scallops sense and retreat from slow-moving dredges, scallop dredges are towed at speeds of up to 2.5 m/s. The scallop dredge has a steel frame with a tongue with an eye, a blade with no teeth, and a bag (Figure 2.1). The width or mouth opening of the dredge ranges from 3.0 to 4.5 m, and dredge weight varies from 500 kg to 1000 kg. The largest scallop dredge vessels, about 60 m long drag two 4.5 m dredges, one from each side of the vessel, and they use winches and navigational electronics to maintain high efficiency. Scallop dredges disturb the seabed. Disturbance of the seabed by dredges is required to dislodge scallops for capture in the net.

Special methods are used to gently extract clams that burrow in sediments. Water jets are directed into the seabed to liquefy the sediment suspending the clams up in a sediment slurry. This blade sieves the clam–sediment slurry. The hydraulic clam dredge is used offshore to collect surf clams and ocean quahogs and inshore to collect soft clams. Hydraulic dredges are efficient, but they restructure sediments and disturb sediment biota.

Offshore vessels (>30 m) slowly tow dredges as wide as 4.5 m across the seabed. The vessels have

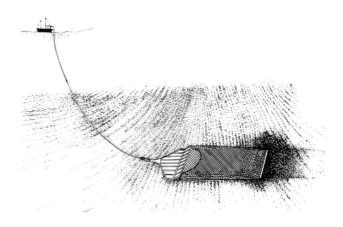

FIGURE 2.1 Action of a scallop and mussel dredge as it is dragged over a soft bottom seabed. The micro-relief of the seabed is smoothed as the dredge is towed over it. The gear in this figure is representative of scallop dredges used in many U.S. fisheries (DeAlteris et al., 1999; used with permission from American Fisheries Society).

large hydraulic pumps that inject water into the sediment through flexible hoses connected to a manifold with multiple nozzles ahead of the blade of the dredge. The dredge is towed slowly so as to not exceed the liquefaction rate. When operated correctly, these dredges are highly efficient, taking as much as 90 percent of the clams in their path.

Management of effort in dredge fisheries is generally achieved with time and area closures and with restrictions on the size (blade width and weight) of dredges, the number of dredges, and the size and horsepower of the towing vessels.

TRAWL GEAR

As fishermen sought to increase efficiency, their vessels became larger and more powerful. This facilitated the evolution of dredges into beam trawls and otter trawls.

Beam Trawls

With the beam trawl, the steel frame used previously in dredges became larger and lighter and the bag became larger and funnel-shaped to concentrate the catch in a cylindrical webbing section, called the codend. The first beam trawls were towed by sailing vessels, but today large beam trawls, with mouth openings of 15–20 m, are towed from both sides of high horsepower trawlers. Beam trawls do not open or sweep areas that are as large as those covered by otter trawls, but are used in selected flatfish fisheries because the beam maintains a constant opening and allows for the attachment of many heavy tickler chains to the frame. The chains disturb the seabed ahead of the sweep of the net to stimulate an escape response in flatfish that have burrowed into the sediment. The gear is towed quickly, up to 2.5 m/s so escape is unlikely and catch efficiency is high.

The frame of the beam trawl consists of an elevated pipe or beam and shoes that tend the seabed at each side. The beam is the site of attachment both for the upper section of the net and for the bridle, which distributes the towing force. The vertical opening of the net is restricted to the height of the beam above the seabed. The sweep of the net is attached to the shoes of the beam, as are tickler chains, if they are used.

Beam trawls are used in flatfish fisheries in northern Europe to catch sole and other groundfish species, and they have been used in experimental fisheries for

monkfish, shrimp, and other demersal species in the United States. A modification of the traditional beam trawl is the plumbstaff trawl used by small gillnet vessels (about 10 m) in Alaska. Beam trawls are sometimes used in scientific sampling because they have fixed mouth openings that allow for accurate determination of sampling areas.

In Florida, roller frame trawls are used to harvest bait shrimp. They are similar to beam trawls in that there is a rectangular, rigid frame that holds the net open. The bottom of the frame supports freely turning, slotted rollers that allow the trawl to move over rough bottoms with less scraping of the seafloor than is typical with otter trawls (Barnette, 2001).

Otter Trawls

Otter trawls developed as fishermen sought to increase the horizontal opening of the trawl mouth, but without the cumbersome rigid beam (Figure 2.2). In the late 1880s, Musgrave invented the otter board, a water-plane device that when used in pairs, each towed from a separate wire, opens the net mouth horizontally and holds the net on the bottom. Initially, all otter boards were connected to the wing ends of the trawl, as they are today in the shrimp trawl fishery. In the 1930s, the Dan Leno gear was developed in France by Vigarnon and Dahl. This gear allowed the otter boards (doors) to be separated from the trawl wing ends using

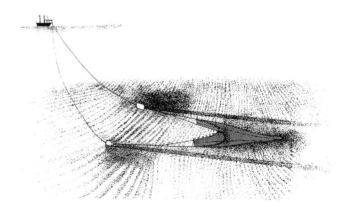

FIGURE 2.2 Action of bottom trawl as it is dragged over a soft-bottom seabed. The otter boards disturb the seabed creating visible tracks and suspending clouds of sediment in their wake. Small-scale sediment relief is smoothed in the path of the trawl net (DeAlteris et al., 1999; used with permission from American Fisheries Society).

cables or "ground gear." This technology increased the effective area swept from the distance between the net wings to the distance between the doors. The ground gear can be as long as 200 m, thus increasing the area swept by the trawl as much as three-fold.

The funnel-shaped bottom trawl net consists of upper and lower sections joined at seams (a twoseam net), called gores. Some fourseam bottom trawls also have side panels to increase vertical opening. The mouth of the trawl net consists of jib and wing sections in the upper and lower panels. A "square" section forms a roof over the net mouth. The body of the trawl net includes belly sections, leading to the codend, where the catch is collected. The webbing is attached to a rope frame consisting of a headrope along the leading edge of the upper panel, and a footrope along the leading edge of the lower panel. The sweep, which tends bottom as the net is towed, is attached to the footrope. The headrope is equipped with floats that open the net mouth vertically. The headrope and footrope or sweep are attached to bridles (also called legs) at the wing ends that lead to the ground wires and the trawl doors.

Box 2.1
Physical Disturbance of the Seabed

Bottom trawl disturbance of the seabed is principally a function of bottom type. On sand and mud bottoms, the trawl door scars consist of small mounds of sediment adjacent to a trough (Figure 2.4). Turbulence created by the passage of the trawl door resuspends fine sediment into the water column, as does the sweep of the trawl net when it scrapes the sand and mud seabed (Figure 2.5). On gravel, cobble, and bedrock bottoms, the trawl doors and the net sweep scrape along the seabed, removing epibenthic organisms and disturbing otherwise stable substrate (Figure 2.6). Small inshore trawlers (15 m), powered by 100+ HP engines, tow small nets (20 m sweep length), have short bridles (10 m) and ground gear (20 m), and have a 20 m span between the trawl doors. Large offshore trawlers (50 m), powered by 1000+ HP engines, tow large nets (50 m sweep length), have long bridles (50 m) and ground gear (200 m), and spread the opening 100 m between the trawl doors. Large vessels with bigger gear sweep larger areas of the seabed (Figure 2.7).

The trawl doors, ground gear, and sweep all disturb the seabed as the gear is towed on the bottom (Box 2.1). The magnitude of the interaction is a function of the design and operation of the specific trawl component and the nature of the seabed. Otter doors range in design from the traditional flat, rectangular boards that create great turbulence in their wake to the modern slotted or foil board that reduces turbulent drag by maximizing hydrodynamic efficiency. Trawl sweeps range in design from simple looped chain sweeps used on smooth, soft bottoms to large roller and rockhopper sweeps used on hard, rough bottoms (Figure 2.3). The development of this more sophisticated trawl sweep technology (rollers and rockhoppers)—combined with sonar technology and the precise navigation afforded by GPS—allows fishermen to trawl areas of the seabed previously considered too rough to fish with mobile gear. However, sonar and GPS also have made it easier for fishermen to avoid areas with rough seabed and thus minimize damage and loss of gear.

There are bottom trawl fisheries for demersal species on all U.S. coasts. In the northeast, 15–50 m vessels fish in waters that are 10–400 m deep. Small mesh nets catch northern shrimp, silver hake, butterfish, and squid. Large mesh trawls catch cod, haddock, flounders, and other large species. Those trawls typically are rigged with long ground wires that create sand clouds on the seabed, herding the fish into the trawl mouth. In the southeast and Gulf Coast areas, small mesh trawls catch shrimp. Because shrimp cannot be herded, shrimp trawl nets are usually connected directly to the trawl doors. Southern shrimp trawl vessels tow two to four trawls from large booms extended from each side of the vessel. On the West Coast, stern trawlers catch shrimp, flatfish, and various gadoid and rockfish species. Factory trawlers or catcher processors, from 50–100 m long, catch, process, and freeze their products onboard.

Pair bottom trawling, once used to catch groundfish in New England waters, is undertaken by two vessels towing a single net. The separation of the towing vessels is used to open the net mouth horizontally. Pair bottom trawls are generally no larger than nets towed by single vessels. The advantage of pair bottom trawling is that considerably more ground gear can be used so as to increase the area swept, due to the reduction in drag resulting from the absence of trawl doors. Use of this gear is now prohibited in the New England groundfish fisheries.

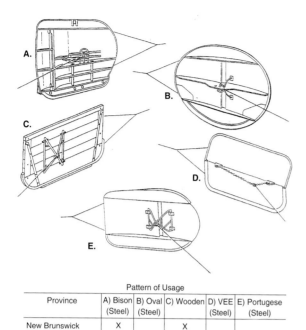

| Pattern of Usage | | | | | |
Province	A) Bison (Steel)	B) Oval (Steel)	C) Wooden	D) VEE (Steel)	E) Portugese (Steel)
New Brunswick	X		X		
Newfoundland		X	X		
Nova Scotia	X		X		
Prince Edward Island			X	X	
Quebec	X	X	X		X

FIGURE 2.3 A) Examples of various trawl door designs used in bottom trawling fisheries. The doors illustrated in a), b), d), and e) are made of steel, and door c) is made of wood (used with permission from Canadian Fishery Consultants Limited).

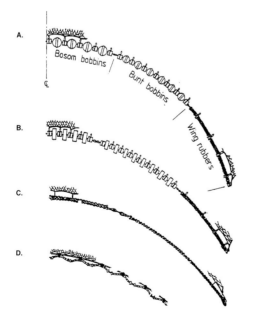

FIGURE 2.3 B) Examples of groundropes used for bottom trawl nets. This figure illustrates a range of the types of groundropes used in trawling gear (used with permission from Canadian Fishery Consultants Limited).

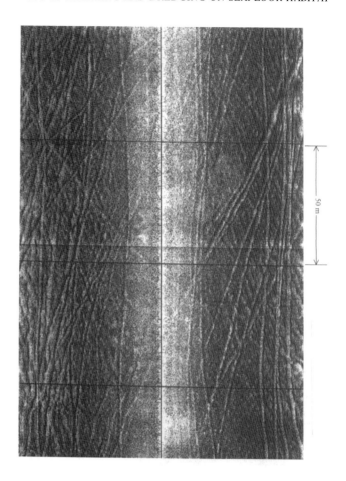

FIGURE 2.4 Side-scan sonar digital image (50 m × 100 m) showing 10+ bottom scars generated by mobile fishing gear in lower Narragansett Bay, Rhode Island (DeAlteris et al., 1999; used with permission from American Fisheries Society).

CONCLUSION

After the passage of the Magnuson-Stevens Act in 1976, the United States promoted the expansion and increased efficiency of the domestic fishing fleet. This included changes in trawl and dredge gear that increased the capacity of fisheries to cover large areas and to reach deeper and rougher habitats. Larger, more powerful boats pulling gear with wider sweeps increased the amount of area potentially affected. Newer gear, such as the raised footrope trawl used in the Gulf of Maine whiting fishery, has been designed specifically to reduce habitat damage.

FIGURE 2.5 Sand cloud generated by a trawl door as it is towed over a flat sand seabed (Main and Sangster, 1981).

FIGURE 2.6 Trawl door on hard seabed showing the absence of a sand cloud pattern. Coarser grained sediments are briefly suspended but rapidly settle (Main and Sangster, 1981).

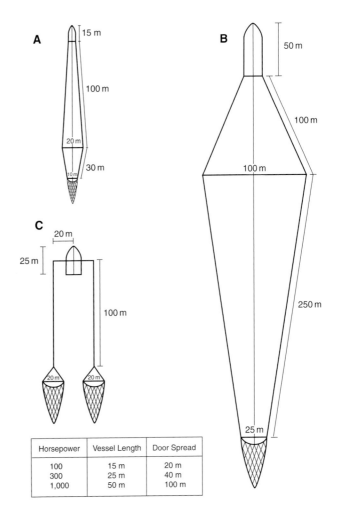

Horsepower	Vessel Length	Door Spread
100	15 m	20 m
300	25 m	40 m
1,000	50 m	100 m

FIGURE 2.7 Scale drawing of representative sizes of trawl gear. This drawing illustrates the relationship between the sweep of the bottom trawl gear and the vessel size.

3

Effects of Trawling and Dredging

"Habitat alteration by the fishing activities themselves is perhaps the least understood of the important environmental effects of fishing" (National Research Council, 1994).

The use of mobile fishing gear has become a source of concern because of the size of the affected fishing grounds, the modification of the substrate, disturbance of benthic communities, and removal of nontarget species. The long-term viability of some fish populations could be threatened if essential fish habitat is degraded. Also, because of declines in many traditional fisheries, efforts to find under-exploited fish populations have increased interest in exploiting less accessible, previously unfished areas. These efforts have been facilitated by the development of new gear and navigational aids. Extensive new regions of the continental shelf, slope, submarine canyons, and seamounts have been exposed to the effects of bottom trawling and dredging. Expansion of fishing into new territory could lead to the loss of habitats that might have provided as refuge for heavily exploited species.

Since the publication of the 1994 National Research Council report, there has been additional research on the effects of fishing gear, especially trawls and dredges, on marine benthic habitats. The magnitude depends on gear configuration, on the subtle modifications various operators make to their gear and on the many and varied habitats fished. Given the inherent difficulty of studying offshore habitats and the problems associated with determining causation under shifting environmental conditions (current, temperature variation, natural migration, storm activity), not all questions regarding the effects of fishing on the sea-floor have been answered—nor are they likely to be in the near future. Evaluating the effects of bottom trawling on benthic communities is complicated by the sparseness of data on species abundance and composition before intensive bottom fishing began. This is important because recent analyses of the few existing historical data sets suggest that larger bodied organisms (fish and benthos) were more prevalent before intensive bottom trawling began (Frid and Clark, 2000; Greenstreet and Hall, 1996). Existing studies necessarily indicate changes relative to recent conditions, not changes relative to the less disturbed ecosystem. There has been an increase in the understanding of fishing gear and habitat interactions that can be used for making decisions about habitat management.

Any fishing gear will affect the flora and fauna of a given location to some degree, but the magnitude and duration of the effect depends on several factors, including gear configuration, towing speed, water depth, and the substrate over which the tow occurs. Variations in substrate include differences in sediment type, bed form (sand waves and ripples, flat mud), and biologic structure (shell, macroalgae, vascular plants, sponges, corals, burrows) (Auster and Langton, 1999). What are the ecologic consequences of these fishing effects? What are the short- and long-term effects on populations, community structure, and interspecific dynamics? Is the disturbance caused by fishing less than what occurs naturally? Are some species threatened with local extinction?

In considering the consequences of trawling and dredging it is important to distinguish between the direct and indirect effects of the activity. Direct effects can be summarized as follows:

- Mortality. Population mortality occurs either as part of the catch (landings plus discards) or incidentally either by killing benthic and demersal species or making them more vulnerable to scavengers and other predators.
- Increased food availability. Discarded fish, fish offal, and dead benthic organisms become food for scavenging species.
- Loss of habitat. Some fishing gears cause the disturbance or destruction of seafloor habitat.

Indirect effects are the downstream consequences of a direct effect. Reductions in the total biomass of target fish, along with the direct effects noted above, could be expected to affect predators, prey, competitors of a target species, and overall seafloor community structure. Indirect effects also encompass potential changes in the flow of materials and energy through ecosystems and shifts in the balance among the processes of primary production, primary consumption, and secondary production.

Human activities such as trawling can be considered a disturbance to environments, and their effects are often compared with natural disturbances that occur in the same ecosystems. It is important to ask whether human disturbances represent selective pressures at novel spatial or temporal scales or are just slight changes in the scale of existing natural disturbance. Natural disturbances can occur with different periodicities, spatial effects, and patterns of recovery (e.g., Lake, 1990; Pickett and White, 1995). Periodic disturbances can be considered pulse events, and a population or community assemblage might respond in several ways. If the disturbance is not too intense, and if the interval between disturbances is long relative to the attributes of the community, or if the system is resilient, the community could return to its previous state. Ecological disturbance theory also suggests that, even if each individual pulse disturbance does not have a large acute effect, there could be a threshold of intensity or a cumulative level beyond which persistent changes in the ecosystem occur. Resilience is the degree to which an ecosystem's long-standing composition, structure, and function can recover from disturbance (Holling, 1973). The disturbance paradigm predicts that short-lived, highly motile or dispersing species with high reproduction rates will recover from disturbance faster than will long-lived, sessile, low-dispersing species (Pickett and White, 1995).

The following sections are based on the results of previous studies and reviews. They summarize com-

monly observed effects of fishing on the seafloor with respect to gear type, the nature of the seafloor habitat, frequency of disturbance (natural and from fishing gear), and rates of recovery to the pretrawling or predredging state.

DIRECT EFFECTS ON SPECIES AND HABITAT STRUCTURE

Research Approaches

Studies of the effects of mobile fishing gear on benthic habitat have used observation and experiment. Observational studies compare the benthic habitat in trawled areas with the habitat in lightly trawled or untrawled places. One difficulty with this approach is finding habitats that are similar in all respects other than the degree of fishing. In any given region, benthic areas inhabited by commercial concentrations of fish and shellfish, not closed by regulation, will be trawled or dredged at some frequency. Quantifying how much trawling has occurred in lightly trawled areas can be impractical given the limited scale of benthic studies. It is difficult to assess how much trawling actually occurs in an area solely from effort data collected in most fisheries. A full evaluation of the effects of trawling and dredging on habitat will require higher resolution effort data to translate the results of small-scale experimental studies to effects at the ecosystem level.

Experimental studies generally use the before/after control/impact design. In this approach, an experimental area is trawled and compared before and after trawling (before/after comparison) and with a site that has not been trawled recently (control/impact comparison). This design often involves direct sampling of fauna, video observations, and sonar scans of the control and disturbed sites. The primary limitation of this design is that it is based on the assumption that the control and experimental sites are equivalent. A study by Lindegarth et al. (2000) suggests multiple evaluation sites are needed to assess the effect of trawling on benthic habitat. The authors showed that the interpretation of experimental studies varies depending on the control and treatment sites compared. Although the need for multiple control sites and replicate trawling is acknowledged within the scientific community, application is limited by ship time, funding constraints, and existing and shifting management regimes.

Research Summary

The effects of mobile bottomfishing gear on benthic habitats depend on the susceptibility of the habitat and on the type gear used. Table 3.1 provides examples of observed effects of different gear in various habitats as catalogued in recent literature reviews (e.g., Auster and Langton, 1999; Barnette, 1999; Jennings and Kaiser, 1988). The extensive primary literature, many review articles, and a meta-analysis of 57 published studies (Collie et al., 2000b), reveal several generalities about the response of seafloor communities to trawling and dredging. These generalities, highlighted in bold text, are discussed below.

Trawling and dredging reduce habitat complexity.
The direct effects of trawling and dredging include loss of erect and sessile epifauna, smoothing of sedimentary bedforms and reduction of bottom roughness, and removal of taxa that produce structure. Trawl gear can crush, bury, or expose marine flora and fauna and reduce structural diversity (Auster and Langton, 1999). On Florida's Oculina Banks, for example, trawl fisheries for rock shrimp and trawl and dredge fisheries for calico scallops have been implicated in the reduction of 1–2 m diameter *Oculina varicosa* tree corals to 2–3 cm rubble (Koenig et al., 2000). If the interval between trawls is shorter than the recovery time, the original benthic structure and species populations might not have the opportunity to recover to pretrawl conditions (Watling and Norse, 1998). Most research bears out the paradigm of variable environments inhabited by short-lived species recovering more rapidly than stable

TABLE 3.1 Examples of Mobile Fishing Gear Effects on Habitat (Based on Reviews by Auster and Langton, 1999, and Barnette, 1999)

Gear	SAV	Sand	Hardbottom/ Biogenic	Muddy Sand	Gravel
Scallop dredge	Increased dredging resulted in significant reductions in biomass and number of shoots (1)	Smoothed bedforms; reduction of epifaunal coverage; shell aggregate dispersal (2, 3, 4)	• Single passage can kill 70% of the living maerl in the dredge path. Flora and megafauna to a depth of 10 cm are damaged. • Dredge tracks remain visible for 2.5 years in maerl habitats. • Maerl is a "living sediment" that is slow to recover from disturbance due to extremely low growth rates (5)	A gradient of increasing large epifaunal cover correlated with decreasing fishing effort (4)	• Undredged sites had higher numbers of organisms, biomass, species richness, and species diversity than dredged sites. Undredged sites had bushy epifauna, dredged sites were dominated by hard-shelled mollusks, crabs, and echinoderms (6, 7) • Suspended fine sediment and buried gravel below the sediment water interface (3) • Smoothed bedforms; hydrozoan cover removed; reduced densities of shrimp (2)
Oyster dredge	Gear modified for clam harvest - reduction in coverage; loss of rhizomes; extended recovery time; sediment suspension; smothering of SAV (8)		Reduction in height of oyster reefs, increased susceptibility to hypoxia (9)		

continued

TABLE 3.1 Continued

Gear	SAV	Sand	Hardbottom/ Biogenic	Muddy Sand	Gravel
Otter trawl	Reduction in coverage; loss of rhizomes; sediment suspension; smothering of SAV (10)	• Reduction of epifaunal coverage; smoothed bedforms; compression of sediments; sediment suspension (fines); reduction in depth of oxygenated sediments (4, 11, 12) • Roller gear produced depressions; chain gear caused damage or loss of epifaunal coverage (11, 13) • Well buried boulders removed and displaced from sediment; trawl doors smoothed sand waves; penetrated seabed 0–40 mm (14)	• Reduced density and size of bryozoan colonies in trawled areas vs. closed areas (15) • Trawled areas showed mussel beds of lower structural complexity and less attached epibenthos compared with untrawled areas (16)	Reduction of epifaunal coverage; smoothed bedforms; compression of sediments; sediment suspension (fines); reduction in depth of oxygenated sediments (4, 14, 17)	
Beam trawl		Trawl removed high number of the hydroid *Tubularia* (17)			50% reduction in density of epifauna such as hydroids and soft coral (18)
Roller-rigged trawl			Damage or loss of sponge and coral cover (11, 19, 20)		• Significant reductions in density of structural components of habitat (21) • No differences in densities of small sponges; 20% of boulders moved or dragged (22)
Roller frame bait shrimp trawl	Minimal SAV degradation; mostly from propeller scars (23, 24)		• Damage, loss of sponge and coral cover (25) • 30–80% damage to coral; declines in groups of large and small benthos in trawled areas (26)		

NOTE: SAV refers to submerged aquatic vegetation.
SOURCE CODE: 1=Fonseca et al., 1984; 2=Auster et al., 1996; 3=Caddy, 1973; 4=Thrush et al., 1998; 5=Hall-Spencer and Moore, 2000; 6=Collie et al., 2000b; 7=Collie et al., 1997; 8=Moore and Orth, 1997; 9=Lenihan and Peterson, 1998; 10=Guillen et al., 1994; 11=Sainsbury et al., 1997; 12=Schwinghamer et al., 1998; 13=Smith et al., 1985; 14=Bridger, 1970; 15=Bradstock and Gordon, 1983; 16=Magorrian, 1996; 17=de Groot, 1984; 18=Kaiser and Spencer, 1996b; 19=Moore and Bullis, 1960; 20=Van Dolah et al., 1987; 21=Engel and Kvitek, 1998; 22=Freese et al., 1999; 23=Futch and Beaumariage, 1965; 24=Meyer et al., 1999; 25=Berkeley et al., 1985; 26=Tilmant, 1979.

communities composed of sessile, long-lived species, which sustain longer term damage.

Repeated trawling and dredging result in discernable changes in benthic communities. Many studies report that repeated trawling and dredging causes a shift from communities dominated by species with relatively large adult body size toward dominance by high abundances of small-bodied organisms (Auster et al., 1996; Engel and Kvitek, 1998; Jennings et al., 2001; Kaiser et al., 2000; Kaiser and Spencer, 1996b; Watling and Norse, 1998). Intensively fished areas are likely to remain permanently altered, inhabited by fauna that readapted to frequent physical disturbance. In some habitats these differences will be profound, in others they will be rather subtle (Kenchington et al., 2001). Species richness (the number of species per unit area) and evenness (the relative abundance of resident species)—two measures of species diversity—can decline in response to bottom fishing, but not all communities show reduced diversity. For example, if bottom fishing reduces the abundance of a dominant species, the disturbed community might have higher evenness and hence lower species diversity (Collie et al., 1997). Untrawled, silty habitat in the Aegean Sea had lower species diversity than did similar, trawled, silty habitat. Measurements of species diversity is not always a reliable indicator of disturbance because a change in the structure of the benthic community can increase or decrease overall species diversity.

Bottom trawling reduces the productivity of benthic habitats. It has been hypothesized that the shift to communities of smaller, fast-growing species after removal of larger, slow-growing species by trawling could maintain benthic productivity and support predacious fish. However, Jennings et al. (2001) found a 75 percent reduction in total infaunal productivity between unfished and heavily trawled areas. Although productivity per unit biomass was higher in heavily trawled areas because of the shift to smaller organisms, overall productivity was lower because of the loss of biomass.

The effects of mobile fishing gear are cumulative, and depend on trawling frequency. Repeated trawling (or dredging) can exceed a threshold above which a disturbance can result in observable, long-term ecological effects. Even shallow, high-energy areas that often experience natural disturbances can be affected if the frequency and seasonality of the trawling disturbances

are different from those of natural events (e.g., Auster and Langton, 1999). Small-scale fishing disturbances can be masked by larger scale natural events (Kaiser and Spencer, 1996b). A three-year study by Tuck et al. (1998) compared the benthic infauna at sites that were trawled regularly and at untrawled control sites. After five months of trawling, they observed changes only in the relative abundances of different species, but after 16 months total species richness began to decline in the trawled sites. Unfortunately, most research has focused on acute effects, quantifying changes to benthic habitat after only a limited number of trawl passes over a short period. These acute studies do not document long-term changes attributable to repeated trawling and dredging. More long-term studies are needed to assess the full range of consequences in areas that are trawled or dredged regularly.

Fauna that live in low natural disturbance regimes are generally more vulnerable to fishing gear disturbance. According to ecologic disturbance theory, initial responses and rates of recovery from trawling should reflect the stability of the substrate in a particular habitat and the character of the benthic community that it supports (Figure 3.1) (Lake, 1990; Pickett and White, 1995). Habitats consisting of unconsolidated sediments that experience high rates of natural disturbance can have more subtle responses to trawling than will habitats characterized by boulders or pebbles (Tuck et al., 2000; Kenchington et al., 2001). Animals that live in unconsolidated sediments in high natural disturbance regimes are adapted to periodic sediment resuspension and smothering like that caused by mobile bottom gear. In contrast, epifaunal communities that stabilize sediments, reef-forming species, or fauna in habitats that experience low rates of natural disturbance have been observed to be particularly vulnerable. Individual studies support the generalizations summarized in Figure 3.1, but a quantitative meta-analysis was less conclusive (Collie et al., 2000a). Responses in sand habitats were usually less negative than in the other habitats, but a consistent ranking of impacts with respect to *a priori* expectations by habitat did not emerge. However, the outcome of the meta-analysis could be confounded by limitations in the available data and by interactions among the factors (gear type, habitat type).

Fishing gears can be ranked according to effects on benthic organisms. Intertidal dredging (with gear that causes the direct removal of sediments, shells, and

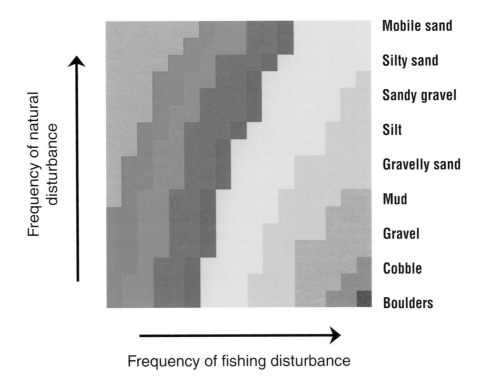

Frequency of natural disturbance

Frequency of fishing disturbance

Mobile sand

Silty sand

Sandy gravel

Silt

Gravelly sand

Mud

Gravel

Cobble

Boulders

FIGURE 3.1 Conceptual model of fishing disturbance to benthic communities. The response variable is the percent decrease in abundance due to bottom fishing. The response is ranked from lowest (top left) to highest (bottom right). The frequency of natural disturbance corresponds roughly with sediment type, but not directly with particle size. The axes correspond to measurements that should be readily obtainable for most parts of the continental shelf.

rocks) has more marked initial effects than either scallop dredging or intertidal raking, which in turn cause greater damage than beam and otter trawling (Collie et al., 2000a). Otter trawls have been evaluated more often than have other types of gear, because of their widespread use (Barnette, 2001; Collie et al., 2000a). This ranking is consistent with the degree of bottom contact and sediment penetration of the different gears.

Benthic fauna can be ranked according to vulnerability. The most consistent research observation is that vulnerability to mobile gear is predicated on the morphology and behavior of the benthic species. Soft-bodied, erect, sessile organisms are more vulnerable to mobile gear than are hard-bodied prostrate organisms. Despite limits in the taxonomic resolution of the data, the meta-analysis identified a 68 percent reduction in anemone abundance, as opposed to a 21 percent mean reduction in starfish, after a single trawling event (Collie et al., 2000a). Similarly, chronic exposure

(repeated dredging) resulted in a 93 percent reduction for anemones, malacostracan crustaceans, brittle stars, and polychaetes, whereas a single dredge event resulted in a 76 percent reduction. On average, none of these taxa increased in abundance, and the average reductions across taxa amounted to 55 percent (Collie et al., 2000a).

Modeling Mortality in Relation to Fishing Effort

Based on the general principles outlined above, a model can be derived to predict the effects of bottom fishing. The depletion of a nontarget species can be modeled with the exponential equation:

$$N_E = N_0 e^{-mE} \qquad [1]$$

N_0 is initial abundance, and N_E is abundance after E passes of a particular kind of fishing gear. The mortality coefficient, m, is analogous to catchability: it includes the mortality of individuals not captured but still killed by the trawl. The mortality rate depends on factors such as gear, habitat type, and life history. One obvious but important implication of this exponential model is that repeated trawls at the same location kill diminishing numbers of organisms. Hence, if the distribution of the nontarget species is not positively correlated with that of the target species, a more aggregated fishery will inflict a lower mortality rate.

The depletion equation can be normalized to the proportion of animals killed as a function of the number of tows:

$$\frac{N_0 - N_E}{N_0} = 1 - e^{-mE} \qquad [2]$$

This response variable is bounded between 0 and 1, and larger values correspond to greater hazard or risk. One can envision a family of curves corresponding to different values of each explanatory variable (Figure 3.2). The curves show that species in sandy habitats experience a lower mortality rate than do those in a gravel habitat. In very few cases have the shapes of these mortality curves been systematically measured. Most trawl studies consist of a single disturbance event (1 tow) or spatial comparisons of chronically fished and unfished areas (at the asymptote of the curve).

The depletion equation also can be expressed as a linear model of potential explanatory variables:

$$\ln\left[\frac{N_E}{N_0}\right] = -(\overline{m} + aG + bH = cD)E \qquad [3]$$

Mortality, m, has been expanded as a linear combination of factors (G, gear; H, habitat) or continuous variables (D, depth). This equation provides the basis for estimating the importance of potential explanatory variables and is similar to the response variable used in the meta-analysis of Collie et al. (2000a).

Modifications to Habitat Structure

An important consequence of trawling is the reduction in habitat complexity (architecture) that accompanies the removal of sessile epifauna and the alteration of physical structure, such as rocks and cobble. Emergent epifauna, such as sponges, hydroids, and bryozoans, provide habitat for invertebrates and fishes. Disturbance of emergent epifauna can increase the predation risk for juvenile fish. Decreased prey abundance increases the foraging time for juvenile fish, thus exposing them to higher predation risk (Walters and Juanes, 1993). Laboratory studies (Lindholm et al.,

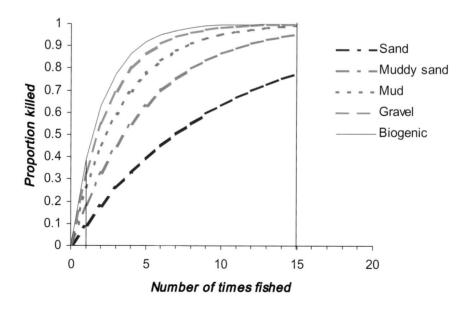

FIGURE 3.2 Hypothetical depletion curves for non-target species in different habitats. The vertical lines indicate that most trawl-impact studies either have been acute (trawled once, vertical line at 1) or compare chronically fished areas (vertical line at 15).

1999) and field studies (Tupper and Boutilier, 1995) have shown that increased epifaunal cover reduces predation risk to juvenile cod.

Information on the linkages between habitat and fish population dynamics is limited; most experimental studies have been conducted in coral reef systems. An extensive literature shows links among larval supply, postsettlement predation, physical attributes of habitat, and adult population size (e.g., Sale, 1991). For example, Sainsbury et al. (1997) provided compelling evidence that loss of structural epibenthos in a tropical system resulted in a shift from a high-value community dominated by *lethrinids* and *lutjanids* (emperors and snappers) to a lower value one dominated by *saurids* and *nemipterids* (lizard fish and bream). By inference, structurally rich habitats in temperate ecosystems also can support a greater diversity of fish species, but the influence of habitat structure on the productivity of economically important species in temperate and boreal ecosystems has not been determined. Where studies have been conducted, and they have been mostly correlative—results are consistent with the assumption that there are linkages between habitat attributes and fish survivorship (e.g., Auster et al., 1995, 1998; Langton et al., 1995; Stein et al., 1992; Tupper and Boutilier, 1995; Yoklavich et al., 2000).

With repeated trawling, the physical relief of the seafloor could be reduced, with a concomitant decrease in the quality of habitat for some species. Juveniles of many demersal fish species are known to aggregate near seabed structure. In trawled areas of the North Sea, the abundance of larger bodied, long-lived benthic species was depleted more than that of smaller, short-lived species, and there was an overall reduction in benthic production (Jennings et al., 2001). Also, removal of physical structure in a habitat can force some species into less optimal environments. For instance, the dredging of oyster reefs in North Carolina has lowered the reefs' vertical height relative to the seafloor. Thus, the only suitable substrate for the oysters is closer to the bottom in deeper areas that are more prone to anoxic events that result from nutrient overloading (Lenihan and Peterson, 1998; Lenihan et al., 2001).

The life histories of demersal fishes exhibit a gradient of linkages to habitat attributes, and the degree of habitat affinity varies by life-history stage. Identifying and quantifying linkages is difficult, especially with data collected during routine population surveys. Figure 3.3 illustrates how the proportion of overall mortality mediated by habitat attributes could change based on life stage and movement rate (as a proxy for

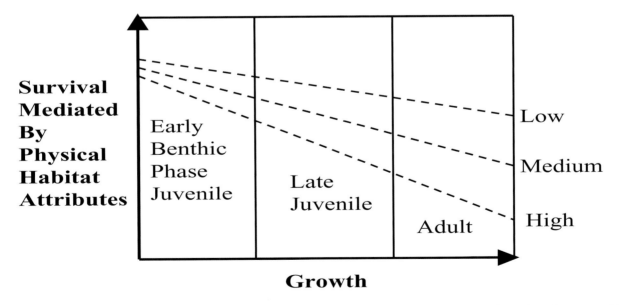

FIGURE 3.3 Conceptual model of the link between habitat attributes and mortality of demersal fishes based on general life history stages. Survival mediated by physical habitat attributes is a direct function of annual movement rates (low, medium, high) that serve as a proxy for habitat affinity. Movement rates are based on movements between habitat patches and are a function of patch size.

habitat affinity). Early benthic-phase juveniles have the highest rates of habitat-mediated mortality. Mortality rates for species that migrate become uncoupled from the physical attributes of habitat because their growth occurs at stages that are earlier than found in species that move less often and have greater affinities for habitat. For example, mortality rates for early benthic-phase Atlantic cod vary in relation to substrate complexity, but adult cod do not seem to exhibit particular small-scale shelter-related behaviors. Mortality of adults can be attributed to disease, senescence, predation from sharks and other large piscivores, and fishing.

INDIRECT EFFECTS

The relatively few studies of the indirect effects of trawling and dredging on marine ecosystems show results that are consistent with the basic principles of marine ecosystem dynamics and predator–prey interactions. Those potential indirect effects, summarized in Box 3.1, should be considered in evaluating the effects of fishing, and they should be used to inform future research and management decisions.

Box 3.1
Potential Indirect Effects

- *Nutrient cycling.* Seafloor trawling and dredging could increase or decrease the exchange rate of nutrients between the sediment and water column and introduce pulses of productivity in addition to pulses from the natural seasonal cycle.
- *Community structure.* Decreased abundance of demersal species could alter trophic linkages. For example, disruption of predator–prey relationships could cause a cascade of changes in other parts of the community.
- *Ecosystem processes.* Trawling and dredging remove ecosystem engineers—organisms that are responsible for water purification (oysters), substrate stabilization, and structure formation.
- *Increased susceptibility to other stressors.* Loss of physical structure in a habitat can expose organisms to other stressors, such as predation and hypoxia.

Sediment Processes

Fishing gear that disturbs the sediment surface can change sediment grain size distribution or characteristics, suspended load, and the magnitude of sediment transport processes (Churchill, 1989; Dyekjaer et al., 1995; Pilskaln et al., 1998; Riemann and Hoffmann, 1991). For example, water jets used in hydraulic dredges to harvest razor clams fluidize substrate for extensive periods (Tuck et al., 2000). Because water content and pore water turnover are important determinants of nutrient regeneration in marine sediments (Hopkinson et al., 1999), hydraulic dredging could alter the nutrient flux.

Bottom trawling and dredging can both resuspend and bury biologically recyclable organic material, changing the flow of nutrients through the food web (Mayer et al., 1991). Studies in relatively shallow depths (30–40 m) show a reduction in primary production by benthic microalgae after a disturbance (Cahoon et al., 1990, 1993; Cahoon and Cooke, 1992). Hence, disturbance in shallow water, including resuspension in the wake of trawls and dredges, could affect nutrient recycling and cause shifts in the abundance or type of microalgae.

The effects of gear-induced disturbance on ecosystem processes are difficult to predict for large marine ecosystems. It could be easier to identify systemwide effects at small spatial scales in semi-enclosed systems—such as bays, estuaries, or fjords—where water exchange with openshelf waters is restricted. However, in open coastal and outer continental shelf systems, the effects of gear disturbances can be small relative to the scale and rate of natural processes. Therefore, the spatial and temporal extent of disturbance by trawl and dredge gear should be evaluated to place these indirect effects within the context of the size and complexity of the ecosystem.

Species Interactions

Direct alterations of habitat can cause species shift and a general decline in the abundance of benthic organisms. Even species that are not directly exploited by a fishery are likely to be affected by the removal or disturbance of benthic and demersal biomass. For example, the early life stages of some pelagic species reside in or depend on benthic communities, for food and shelter. It is difficult to separate the indirect effects of trawling and dredging on benthic and pelagic communities from other sources of variation, such as climate change (Jennings et al., 2001).

In some cases, removal of one species can have cascading effects on the rest of the ecosystem. For example, the combination of disease and high harvest rates over the past 150 years has reduced oyster density in the Chesapeake Bay to less than 1 percent. The loss of the filter-feeding oyster's capacity to consume algae is hypothesized to be partially responsible for the proliferation of algal blooms. This appears to have shifted the composition of the pelagic community from mesozooplankton and fish to a community dominated by predatory jellyfish and comb jellies (Caddy, 1993; Ulanowicz and Tuttle, 1992).

RATES OF RECOVERY

Recovery is the return of an ecosystem to a state that existed before a disturbance, as measured by ecosystem processes, species composition, and species interactions. Recovery from trawling will depend on the type and extent of the habitat alteration, the frequency of the disturbance compared with natural changes, habitat characteristics, and species and life history characteristics. Recovery times vary according to the intensity and frequency of the disturbance, the spatial scale of the disturbance, and the physical characteristics of the habitat (sediment type, hydrodynamics). Superimposed on these human-related alterations are natural fluctuations, caused by storms or long-term climate changes, for example.

In most circumstances, only a first-order approximation of recovery rate is possible. Experimental evaluations recovery after cessation of trawling are limited and have focused on biotic recovery of small-bodied, short-lived invertebrates. Despite that, we can make some observations about the amount of physical disturbance that is sustainable in some types of habitat. The meta-analysis by Collie et al. (2000a) showed that recovery rate appears to be slowest in the more stable muddy habitats and biogenic (structurally complex) habitats. By comparison, mobile sandy sediment communities could be able to withstand 2–3 trawl passes per year without changing markedly. It is important to bear in mind, however, that although available data allow for prediction of the recovery rate for small-bodied taxa such as polychaetes (which dominate data sets for sandy sediment communities), less abundant, long-lived, and hence more vulnerable species could recover more slowly.

In some biogenic habitats, physical disturbance by dredging and trawling has a long-lasting effect. For example, clam dredging causes severe and persistent changes to seagrass ecosystems (Peterson et al., 1987; Stephan et al., 2000). After a single pass, seagrass biomass fell by about 65 percent below controls, and recovery did not begin for more than two years with seagrass biomass still roughly 35 percent below controls four years later (Peterson et al., 1987). The abundance of fish and shellfish that depend on seagrass for settling locations for protection from predators could be reduced where seagrass is damaged.

Environmental recovery after disturbance depends on the life histories of the organisms that live in or create the habitat. Recovery time is often one to five times the generation time of the organism (Emeis et al., 2001). Therefore recovery times could range from a few months—or less—to several decades (Hutchings, 2000). Many of the larger biogenic structure-forming organisms, such as soft corals and sponges, are slow growing and long-lived (Dayton, 1979; Leys and Lauzon, 1998). Empirical data about recovery times of corals and coral-line algae are sparse, but recovery times of decades to centuries can be inferred from the age of these organisms.

Recovery from trawling also depends on the size of the area disturbed (Thrush et al., 1998), and on the spatial pattern of the disturbance (Auster and Langton, 1999). Each trawl track is a small disturbance, but over a long enough period and with widespread coverage, the small changes can result in a large effect. The consequent habitat loss, and effects on resident species, depends to a large extent on its scale (Deegan and Buchsbaum, 2002). A single small loss might not, by itself, have an observable effect on species that are not directly damaged by trawling. However, the cumulative impact of many small losses may be quite significant at a regional scale (Odum, 1982). In some coastal ecosystems, mosaic-type damage could allow faster recovery than would a large-scale, isolated disturbance (Emeis et al., 2001).

Areas that are trawled with greater frequency could take longer to recover. Almost all studies have examined recovery after a single, acute pass by a trawl rather than after the multiple passes that are typical in frequently trawled, heavily fished areas. There, recovery would be expected to take longer because a larger fraction of the population is removed and immigration rates are lower (Figure 3.2). Results from the meta-analysis (Collie et al., 2000a) indicate that, on average, a single dredge event results in a 76 percent, whereas repeated dredging results in an average reduction of 93 percent for anemones, malacostracan crustaceans, brittle stars, and polychaetes (Figure 3.3).

Few studies have examined the recovery of ecosystem processes or whole communities. Brylinski et al. (1994) showed that trawling significantly affected benthic diatoms that occurred in the intertidal zone, but that recovery occurred at all stations after about 30 days. The higher light intensity (and spectral composition) in the experimental area than at deeper sites, where trawling normally occurs, might have contributed to the relatively fast recovery.

The limited findings to date concur with theoretical predictions that suggest longer recovery times for more stable and complex habitats (Auster, 1998; Auster and Langton, 1999; Kaiser, 1998). Clearly, habitats with extended recovery periods are strong candidates for protection from disturbance caused by fishing. However, much better data on the geographic distribution and long-term effects of chronic trawling and dredging in these physically more stable habitats are required to estimate recovery rates that will promote strategic, rather than precautionary, management decisions.

UNCERTAINTY

Underlying the concept of the reversibility of the effects of dredging and trawling is the implicit assumption that eventual recovery to the former state will occur if the activity is halted. This assumption derives from an ecological paradigm in which ecosystems and communities are viewed as part of a successional continuum along a disturbance gradient. An alternative approach recognizes multiple-equilibria, non-linearity, and threshold effects (Holling, 1973; Holling et al., 1995; Patten and Constanza, 1997). In an alternative state, ecosystems have different species compositions, functions, and ability to provide ecological services. They might therefore be valued quite differently by society. Resilience, or the counteractive capacity of the ecosystem, is measured by the ability to maintain structure and function in the presence of stress or disturbance. When resilience is exceeded, the system can flip to an alternative state from which it will not return simply by removing the source of disturbance (Holling, 1973; Holling et al., 1995). These regime shifts can affect valuable ecosystem services, including fisheries yield (Collie and Spencer, 1994; Knowlton, 1992).

Human or natural modification of the marine environment might result in the shift of a community from one stable state that provides economically valuable fish to another stable state dominated by fish of higher or lesser value. Ecosystems respond to perturbation in many ways, including changes in species composition (by loss, inclusion, or replacement) and in the relative abundance of biomass (with an increase or decrease) of some species. Overall production and biomass of an ecosystem can remain the same as species respond to natural or human-induced stress (some species increase while others decrease) (Breitburg, 1998; Fogarty and Murawski, 1998). This change to a new stable state might take place abruptly, right after the disturbance occurs, or it could result from small cumulative shifts in natural forcing variables. Examples of species replacements are the apparent increase in cephalopod species in the Gulf of Thailand, which coincided with the increase in trawl fishing and the reduced abundance of demersal fish, and the increase in pelagic species that seems to have occurred in the North Sea and elsewhere. Its duration may vary. The return time to the initial stage has been predictable in some cases. In other cases, with nonlinear interactions and multiple-equilibrium states, the time the ecosystem will remain in a new state is not predictable. For benthic communities with a long history of fishing disturbance, it is unknown whether the community would return to the undisturbed state if the disturbance were stopped.

Human modifications to marine environments compromise the capacity of marine populations to recover from stresses, such as storms, eutrophication, and climate change, whether natural or anthropogenic. Seagrass ecosystems provide an example in which the synergistic effects of habitat loss due to trawling could compromise the ability of the system to withstand or recover from other disturbance. Seagrass ecosystems are important habitats and locations of fisheries for numerous fish and invertebrate species. The natural distribution of seagrass habitat is controlled by light availability that is a function of water quality, including the presence of phytoplankton and suspended sediments. The physical structure of seagrass, including stem density and the size of beds, increases water clarity by filtering water column particulates and depositing them on the bottom (Thayer et al., 1984). This creates a zone of clear water around seagrass beds that allows them to persist and expand. Trawling can fragment the seagrass bed into small pieces that do not effectively trap suspended particles, resulting in light limitation. Eutrophication also enhances the proliferation of faster growing phytoplankton, epiphytic algae, and macroalgae that compete with seagrass for light and space (Kemp et al., 1983; Phillips et al., 1978; Short et al., 1995; Twilley et al., 1985). Light limitation of seagrasses leads to diminished growth and stature, increased shoot mortality and declines in shoot density (Duarte, 1995; Moore et al., 1996; Short et al., 1995), resulting in declines in seagrass habitat area. Initial habitat fragmentation by trawling and dredging can make seagrass habitats more susceptible to the negative effects of eutrophication.

The maintenance of the ecosystem in an alternative state will depend on interactions with adjacent ecosystems and the intensity of the new biologic links. Additional disturbance generated by natural events or by new trawling and dredging can help maintain the assemblage in this state of equilibrium or transfer it to a new state. In the benthos, disturbances can be physical (hurricanes, suspension of sediment by surf, lateral transport by bottom currents, seasonal hypoxia generated by the input of nutrients, limited export of biogenic carbon) or biological (predation, flux and export of biogenic carbon, deposition of debris, bioturbation, competitive exclusion). Their common action is to remove organisms and to open spaces for colonization by other organisms. If disturbances are frequent, gaps will constantly reset to one of the multiple stable stages. If disturbances are rare, most of the community will remain in a stable state for most of the time. The loss of complexity and biodiversity can threaten important ecologic functions (the cycling of important elements or the control of populations of particular species) or the resilience of ecosystems to change or disturbance.

SUMMARY

For the most part, existing information about the direct responses of benthic communities to trawling and dredging is consistent with the general principles that govern how ecologists expect communities and ecosystems to respond to acute and chronic physical disturbance. Trawling and dredging change the physical habitat and biologic structure of ecosystems and

therefore can have potentially wide-ranging consequences. Mobile gear reduces benthic habitat complexity by removing or damaging the actual physical structure of the seafloor, and it causes changes in species composition. The reduction of physical structure in repeatedly trawled areas results in lower overall biodiversity. Of direct concern to commercial and recreational fisheries is the possibility that losses of benthic structural complexity and shifts in community composition will compromise the survival of economically important demersal fishes. Mobile gear also can change surficial sediments and sediment organic matter, thereby affecting the availability of organic matter for microbial food webs.

RESEARCH NEEDS

It is clear that the links between habitat alteration and loss of fisheries production can be subtle and diverse and that they operate on many spatial scales, from site-specific to regional. Most studies have been done in shallow water in small areas. Researchers have examined acute disturbances, rather than chronic, and they have studied short-term response focused in animal communities, as opposed to ecosystem processes such as nutrient regeneration. Although there have been many acute studies, few have examined the effects of short-term multiple passes, and future research should address this type of disturbance.

Perhaps the biggest research gap is on chronic effects and recovery dynamics. More studies on chronic disturbance by fishing gear are needed to determine the dose–response relationship as a function of gear, return time, and habitat type. Research also should address recovery dynamics, with consideration given to estimating the large-scale effects at current fishing intensities (e.g., Collie et al., 1997). This research should include quantitative studies undertaken in deeper water (>100 m) and studies in stable and structurally complex habitats, for which the recovery trajectory will be measured in years to decades. The statistical power to detect fishing effects will be greatest when biologic sampling can be combined with high-resolution spatial data on fishing effort.

4

Habitat Mapping and Distribution of Fishing Effort

Chapter 3 describes the remarkable similarity across studies of the observed effects from fishing with mobile gear. Most studies have covered small areas and have focused on acute effects, but a few studies have addressed chronic effects, and they have found the same patterns (reduced habitat complexity, shifts in community composition, reduced diversity). More information is needed before the small-scale response patterns can be used to make quantitative estimates of ecosystem effects and recovery times. Two fundamental data needs for scaling the observed effects of trawling and dredging on marine habitats to the ecosystem level are the type and magnitude of the effects of specific gear on different habitats, and the spatial and temporal extent of fishing activity.

Although precautionary approaches can be implemented using existing data, the goal is to have sufficient information to support tactical decisionmaking. In most areas, mapping of habitats and fishing effort has been done at a relatively large scale compared to the smaller scale at which ecological effects are described. Dredge and trawl fishing grounds cover hundreds of square kilometers, but most study areas are <1 km². Because of that spatial mismatch, estimates based on existing data could either overestimate or underestimate the actual effects of mobile fishing gear.

HABITAT

In Chapter 1, habitat is defined as the environment necessary to support, directly or indirectly, the life processes of the resident organisms. Fish require a broad diversity of habitat processes to survive, grow and reproduce. Typically, physical features have been used as a primary criterion for classifying habitats (Allee et al., 2000) and for characterizing the locale of many research studies on the effects of fishing (Chapter 3). But biologic features, including the composition of biologic communities, are equally important when evaluating the effect of altered habitats on fish populations.

Fishing with bottom trawls or dredges takes place with varying intensity in state and federal waters adjacent to every U.S. coastal state. Most bottom fishing takes place on the continental shelf and upper slope in water depths of less than 500 m, although some extends as deep as 2000 m. The seafloor affected, or potentially affected, by mobile bottom-tending fishing gear includes a wide range of habitats, from relatively featureless sandy and muddy bottoms to highly structured seagrass beds and coral reefs. Because the effects of trawling and dredging are not the same in all habitats, it is essential to know the spatial distribution of habitat types in areas where bottom fishing occurs.

HABITAT CLASSIFICATION

Classification of habitat requires a system that uses common terminology and is adopted uniformly by federal, state, and local agencies. A logical and consistent classification of the environment provides the basis for evaluating the extent and significance of disturbance in each habitat type. There is a great need for habitat maps for fishery management, but no systematic maps of large regions currently exist. Some moderately large areas (several hundred square miles) have been mapped recently (see for example, Reynolds et al., 2001; Valentine et al., 2001; Wakefield et al., 1998) but even these are small relative to the areas involved in many fishery management decisions. In general, habitat maps have been compiled only on an *ad hoc* basis for small areas. This is due partly to the lack of an accepted classification scheme for seafloor habitats in the United States.

Several classification systems have been proposed (Allee et al., 2000; International Council for the Exploration of the Sea, 2001; Roff and Taylor, 2000). These systems are hierarchical: they start at a large scale (1000 km) consisting of permanent physical features and scale down to microhabitat. Each system includes enduring physical features of the environment, such as bottom relief, substrate, temperature, stratification, and exposure. Natural physical features are relatively stable over time and can be measured with broad-scale surveys. Physical features alone, however, do not define habitat. Biologic features, the presence of seagrass beds, kelp forests, and coral reefs, for example, also define habitats.

World Wildlife Fund Canada has sought to apply the concepts of terrestrial habitat mapping to Canadian marine waters (Roff and Taylor, 2000). The Canadian classification scheme is based on enduring physical features of the environment. It is especially important to use physical features compatible with broad-scale surveys in Canada and the United States because of the very long coastlines. The rationale is to delineate relationships between biologic community composition and physical variables that can be measured more readily over large areas. This World Wildlife Fund approach has been used to identify representative and distinctive habitats off the coast of Nova Scotia, and it is being used by the Conservation Law Foundation to characterize habitats in the Gulf of Maine.

A parallel approach has been taken to map and classify marine habitats in European waters. The European Nature Information System (EUNIS) has been developed for the European Environment Agency (International Council for the Exploration of the Sea, 2001). European marine habitats are defined by geographic, abiotic, and biotic features, whether entirely natural or partially modified. This broader definition recognizes that habitats can be defined by biogenic features and that marine habitats have been shaped by human activities (e.g., artificial reefs and shipwrecks). The EUNIS classification has five hierarchical levels (Box 4.1).

The difference between the Canadian and European

**Box 4.1
European Nature Information System Classification**

The European Environmental Agency has been developing a common parameter habitat classification framework for marine and terrestrial systems. This forms an integral part of EUNIS. The marine component was derived from the BioMar project (Connor et al., 1997) and has five basic levels, based on the following criteria: 1) marine and coastal habitats versus freshwater and terrestrial habitats, 2) depth zone, 3) substrate type, 4) biologic features, and 5) dominant species.

Because of differences in biotic communities, the ecotypes used in Levels 4 and 5 are not broad enough to account for the diversity of marine habitats present in U.S. waters. The general features of this classification system are consistent with those developed for the United States (Allee et al., 2000).

---

approaches is largely in available data. In part because of the more uniform geography of European waters, habitats have been mapped more thoroughly, allowing more detailed classification. The U.S. situation is more similar to Canada because the United States has an extensive coastline that encompasses an even greater variety of habitat types. Mapping the entire U.S. exclusive economic zone even by substrate type (EUNIS Level 3) is not possible with existing data. Habitats in the U.S. exclusive economic zone range from arctic to tropical, making it difficult to construct a standardized classification scheme for all habitats. Spatial management is at the km^2 scale and requires mapping of sediment type (EUNIS Level 3). Although the ecological effects of bottom fishing have been documented at the scale of square meters (EUNIS Levels 4 and 5) concerns about anthropogenic effects are at the much larger scales of ecological communities. Therefore, the International Council for the Exploration of the Sea Working Group on Ecosystem Effects of Fishing Activities recommended identification of habitats influenced by human activity and inclusion of biologic characteristics in the EUNIS classification system.

In the United States, two relatively broad marine

habitat classification schemes have been proposed. Greene et al. (1999) proposed a system based heavily on geophysical remote sensing and involving a hierarchy of habitat feature scales. The system does not address shallow marine habitats (<30 m). It was developed specifically for rockfish habitats in 30–300 m of water along the west coast of North America, but could be adapted for characterizing other habitats. Recently, a committee sponsored by the Ecological Society of America and the National Oceanic and Atmospheric Administration's Office of Habitat Conservation (Allee et al., 2000) developed a comprehensive marine habitat classification system for U.S. waters. This hierarchical system encompasses all marine habitats and attempts to allow practical classification of habitats at gross to fine scales, depending on the resolution of the available data. Although broadly similar to the EUNIS system (Allee et al., 2000), the new system considers many more habitat types. It differentiates photic from aphotic environments, considers water column habitats, makes a distinction between marine habitats that are associated with a continental mass and those that are not, and emphasizes depth zones and exposure to wind and wave energy. Anthropogenic disturbances are taken into account using "local modifiers" at the finest level of the classification. Because this classification system is relatively new, it has not received much attention. But it is a promising beginning for the wide variety of marine habitats in U.S. waters, although the system needs to be tested, discussed, and modified by the broader scientific and management community.

The next three sections describe how benthic habitats can be characterized by bathymetry, geologic structures, and biogenic substrates (i.e., the plant and animal structure). This is followed by a description of the value of a geographic information system (GIS) for assembling habitat information and by a discussion of the importance of scale in defining and mapping habitat. Appendix C gives a brief discussion of tools used in mapping habitats.

BATHYMETRY

The basic seafloor map is a water-depth or bathymetry map, which shows the shape of the seafloor. These usually are made acoustically, from individual soundings, continuous lines of soundings, or wide swaths of soundings. A map can display the bathymetric information as numbers at the locations of each sounding, as contours delineating zones of equal depth, or as a

shaded-relief image. The bathymetry of virtually the entire shelf of the U.S. has been mapped at a scale of 1:250,000 with 2 m contours (National Geophysical Data Center, 2002). Locally, particularly in harbors and approaches, greater detail is available. In a few areas, particularly where there are large recreational fisheries, rudimentary substrate maps (National Geophysical Data Center, 2002), which show bottom type overlain on the bathymetry, have been published at a scale of 1:100,000 and a contour interval of 2 m. Higher resolution maps also have been made in many estuaries, bays, and nearshore areas by various federal and state agencies, academic institutions, and private companies with local interests (such as mining, pollution studies, coastal erosion research or mitigation). There are detailed bathymetric maps for some offshore areas of special interest, such as the National Marine Sanctuaries (Figures 4.1 and 4.2). However, for most of the continental shelf where most federally regulated fishing activity occurs, bathymetric maps with 2 m contours offer the highest resolution available. This scale is insufficient for some fishery management purposes because some features (e.g., ledges, boulders, depressions) with relief much less than 2 m can be significant habitats for fishes.

GEOLOGIC STRUCTURE AND SUBSTRATE

The geologic character of the seafloor (texture, morphology, composition, thickness of sediment substrates; distribution of hard bottom; lithology of rock outcrops) defines marine habitats and influence what organisms will live in an area. For example, sessile organisms cannot attach to a fine-grained mud substrate, and burrowing organisms might not be able to maintain their burrows in loose sand. Karstic limestone, with its craggy, dissolved formations, provides ample hiding places for organisms of many sizes; massive quartz sandstone beds might not. Mapping the geologic character of the seabed requires a suite of data types. Sidescan-sonar imagery can suggest the areal extent of bottom types, but samples of the sediments and rocks are necessary to determine what the bottom types are and to analyze the texture and composition of the substrate. Seismic-reflection sub-bottom profiles are used to assess the thickness of sedimentary deposits and the geometry of subsurface strata.

Several large databases of marine sediment texture are maintained by federal agencies (U.S. Geological Survey, Minerals Management Service, and National

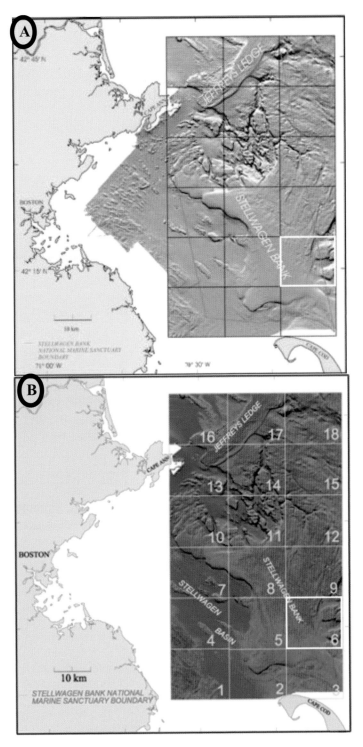

FIGURE 4.1 Bathymetry and backscatter maps from multibeam-sonar data. Multibeam mapping tools make it possible to collect high-resolution bathymetric data over large areas of the seafloor more rapidly than was possible in the past with older single beam systems. The data shown here cover approximately 1,300 square nautical miles of seafloor off Boston, Massachusetts and took about 55 24-hour days of survey time to complete. (A) displays the morphology of the seafloor as though it were illuminated by a light shining down from the north. (B) displays the intensity of the signal that is reflected back to the instrument. Higher intensities are shown in red and correspond to more reflective sediments such as sand and gravel or rocky outcrops. Blue represents lower backscatter and corresponds to less-reflective fine-grained sediments, such as silt and clay. The area labeled "6" is shown expanded in Figure 4.2 (modified from Valentine et al., 2001 and Butman et al., in press).

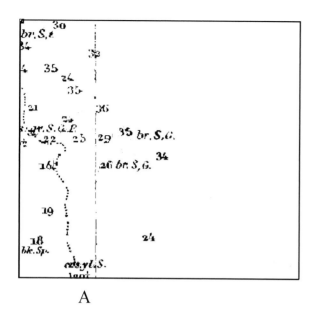

A

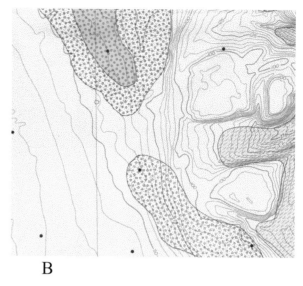

B

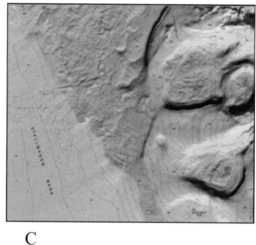

C

FIGURE 4.2 Progression of knowledge of seafloor at Stellwagen Bank, Area 6 (see Figure 4.1 for location). (A) shows the widely spaced soundings obtained by lead line in 1855, when Stellwagen Bank was discovered. Annotations are also made concerning the substrate type (e.g., S for sand, G for gravel). Sediment samples were probably obtained by putting wax on the lead weight which was lowered to measure the water depth. The bathymetry shown in (B) is based on closely spaced lines of acoustic soundings (Schlee et al., 1973). Substrate type is shown as colored areas but much interpolation between sediment samples (black dots) has been done. (C) shows the same area, with highly detailed multibeam bathymetry and the substrate types shown in color are based on a combination of numerous sediment samples, sidescan-sonar data, and bottom photo and video data (Valentine et al., in press).

Geophysical Data Center). Some also include composition and other descriptive or analytical data. In some areas, the databases can provide a large-scale estimate of substrate types, but the distribution of existing sediment samples is highly variable. Some areas, such as the Atlantic continental shelf, have been systematically sampled approximately every 10 nautical miles (Poppe and Polloni, 2000). Smaller areas such as Long Island Sound, can have a mile or less between samples (Figure 4.3). Other areas, such as large parts of the Bering Shelf, have not been sampled at all. It should be noted that even for areas with relatively closely spaced

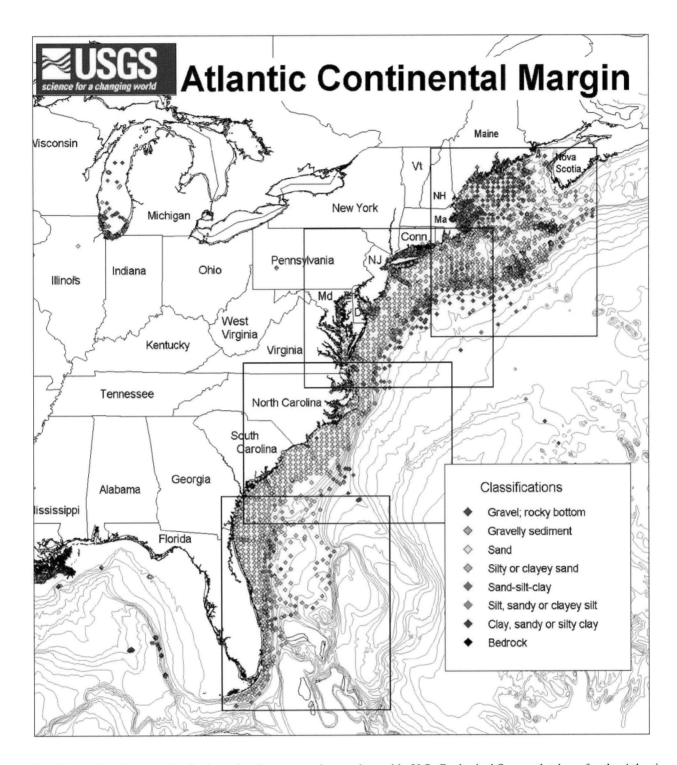

FIGURE 4.3 Map showing distribution of sediment samples catalogued in U.S. Geological Survey database for the Atlantic Continental Margin (Poppe and Polloni, 2000).

samples, the sampling density may not adequately characterize the bottom. Gravel substrate, for example, is more difficult to sample than finer-grained sediments, so it is often underrepresented and sediment texture says nothing about hardbottom or rock outcrops.

Many smaller studies of specific areas have been carried out by federal and state agencies and academic institutions. As is the case with bathymetric data, greater detail is available for nearshore areas, bays, and estuaries; less for the continental shelf; and very little for the continental slope.

GEOGRAPHIC INFORMATION SYSTEM

Significant amounts of data exist that are useful for fishery management, such as the distribution and frequency of trawling, bathymetry, and substrate composition and texture. These data reside in a variety of places, are not always easy to access, and often are highly variable both within and between regions.

GIS has become the standard way to represent geospatial data (i.e., any data that has a position associated with it). GIS does digitally what scientists have long done with tracing paper or clear acetate—it allows layers of information to be overlaid in various combinations so that relationships between the different data sets can be seen. Many types of data useful to fishery management are appropriate for use in GIS: bathymetry, substrate composition or texture, habitat types, current speed and direction, catch and effort data, species distributions, and bottom photographs. Because GIS allows the user to expand or contract the field of view, data of various resolutions can be used together more effectively. This is very important in fishery management because the resolution of available data varies widely and because, as discussed above, different resolutions might be appropriate to address different issues.

Because the marine waters of the United States include a wide range of habitats, fishery management issues differ regionally. Differences in the seafloor geology and regional climate result in different species assemblages and lead to different fishing practices. Specific management questions require higher resolution data of one type in one area and lower resolution elsewhere. In an ideal world, we might create uniform, high-resolution habitat maps of all U.S. continental shelf areas, but limitations of technology, time, and funding make that impractical, if not impossible.

This dilemma can be addressed by constructing a standard system for habitat classification and by compiling all relevant existing data into a single, readily available GIS format. This will allow managers to make the best use of existing data and to see where data gaps exist. Fishing effort data superimposed on habitat maps, for example, would help identify regions where more study is needed or where problems could arise. Presentation of data sets of differing scales is not a problem in GIS: The user can zoom in where the data density warrants a close-up view or zoom out for a regional view. As new data become available they can be added.

QUESTIONS OF SCALE

The scale at which an area is studied is critical to the eventual utility of the outcomes (products, maps, data) of the study. For example, an issue involving larvae in benthic infaunal structures must be addressed with more detailed information than does one involving schools of fish that aggregate in a broad depth range. Conversely, sediment samples to "ground truth" kilometer-wide areas of differing backscatter in a sidescan mosaic need not be spaced every 10 m. In choosing an appropriate scale for mapping, three interrelated issues must be addressed:

1. What is the purpose of the map? What questions are you trying to answer? What attributes of habitat influence the distribution and abundance of the species of interest? Do you need to know precise locations of objects (e.g., gravel, coral, artificial reefs) and if so, how big are they? Would it suffice to know that an area is composed of gravel, and know the extent of that area, but not whether the gravel is mostly boulders, cobbles or pebbles?
2. How homogeneous, or patchy, are the things to be mapped? This might not be known before you begin to map, but usually some indication—anecdotal information, a detailed map of a small portion of the area to be mapped, a few bottom photographs—is available or can be obtained before a commitment to a mapping plan is made.
3. Is the technology or approach to be used appropriate to the scale of information needed? There is no one right way to map the seafloor. If you need to know about hazards to navigation, you must have bathymetry. If you need to know the distribution of gravel substrate, bathymetry alone will be of little help. Sidescan or multibeam back-

scatter data and bottom sampling would give this information. If you also need to know the abundance of particular sessile organisms that form important components of habitat (sponges, coral), you will need to add photographic or video data.

Obviously, the scale needed for various data types will vary from region to region and from issue to issue. Characteristics of the substrate are more homogeneous in some areas than in others.

DISTRIBUTION AND INTENSITY OF FISHING EFFORT

The second component of geographic-based information required for assessing the effects of trawling and dredging is the spatial distribution and intensity of fishing effort. As described earlier, potential changes at the ecosystem level caused by trawling or dredging are a function of the extent and distribution of fishing activities in various habitats.

To give some appreciation of the scale of trawling and dredging effects in terms of frequency and area covered, rough approximations of the relative intensity of effort in the fishery management regions are presented in Table 4.1. These estimates are based on data provided by the various state and federal fishery management agencies and make several assumptions regarding the areas swept, the estimated total fishing area, and the total number of trawl or dredge tows, as described below.

There is considerable regional variation in the collection of data on bottom trawling and dredging activities conducted on the continental shelf and slope off the U.S. Coast. Depending on the region and the fishery this information is collected either by the National Marine Fisheries Service or by state agencies. During the latter part of the past decade, trawl fishing effort was documented in statistical blocks that encompassed about 231,200 nautical square miles of the continental shelf and slope. A statistical block (or reporting region) defines a geographic area that is used for reporting the location and duration of individual trawl tows during a given time period. The statistical block may be based on a grid or other feature, and often varies in size among regions (Table 4.1). Trawl fishing effort, as monitored by various national and international fisheries agencies, is normally defined as the number and duration of tows (drags) made in a particular area (statistical block or region), over a specified

period. Effort data is frequently reported as the number of tows per area, along with information on tow duration. In the Gulf of Mexico and along the northeastern Atlantic seaboard, the trawl effort is reported as the number of 24-hour fishing days (excluding steaming time). In the Pacific regions the number of tows and the associated towing time is logged. **The regional differences in gear and in effort reporting methods and the variation in the geographic resolution of the effort data make exact comparisons among regions unrealistic.** Furthermore, the estimates in Table 4.1 of the total number of tows have been extrapolated in all areas from observer data, log books, shore side vessel sampling, and other sources.

For a given area or region it is nevertheless possible to calculate the potential area swept by dragged gear using the sum of the approximate area covered by individual tows. This calculation provides an upper estimate of the coverage of a reporting area by trawl or dredge gear, or assuming that the individual tows are evenly distributed throughout the area. However, some tows in a statistical block may overlap or be concentrated in one portion of the reporting area (Figure 4.4). If tows are clustered in a fraction of the statistical block, the swept area will be smaller, but the intensity of effort will be higher in the subarea where fishing is concentrated.

Because effort data have not been collected continuously, it is impossible to compare regions for any given year. However, data collection at various times during the 1990s for four of the six fishery management regions with significant trawl or dredge fisheries has allowed the trawl and dredge effort to be mapped by statistical reporting areas for that period. Appendix B contains detailed descriptions and maps of the bottom trawl and dredge effort data for major fisheries. The swept area estimates in Table 4.1 are based on the intensity of effort in defined statistical areas in the fishery management regions based on available data. Even though the existing effort data precludes precise comparisons, estimates of the frequency of effort per unit area suggest that, at times during the 1990s, the highest intensity of trawling occurred in the Gulf of Mexico and New England.

In response to the collapse of the groundfish stocks in New England, managers instituted effort controls and closed three large areas to trawl and dredge gear in 1994. In 1999, a fraction of the closed areas were opened for limited scallop dredging when surveys indicated that the closed areas contained abundant beds

TABLE 4.1 Estimated Fishing Density by Region

Region	Total Statistical Blocks Fished	Block Area (nm^2)	Total Area Fished (TAF) (nm^2)	Total Tows	Swept Area (nm^2)	Percentage TAF Swept/ Year	Percentage of Statistical Blocks Swept More Than Once/Year	Gear Type	Observed Years
New England[a]	561	71.7	40,168	NA	46,193	115	56	bottom trawl	1993
Mid-Atlantic[b]	44	704.7	31,007	NA	11,925	38	5	bottom trawl	1985
Southeast U.S.	NA	NA	NA	NA	NA	NA	NA	NA	NA
Gulf of Mexico[c]	210	variable	78,629	902,885	200,588	255	57	shrimp trawl	1998–1999
Alaska									
Bering Sea[d]	3,791	7.3	27,632	17,688	15,724	57	<2	bottom trawl	1998–2000
Aleutian Islands[e]	711	7.3	5,168	3,650	2,974	58	3	bottom trawl	1998–2000
Gulf of Alaska[f]	1,553	7.3	11,320	8,640	5,120	45	1	bottom trawl	1998–2000
West Coast									
California[g]	264	78.3	20,671	15,535	6,902	33	14	bottom trawl	1994–1996
Oregon & Washington[g]	373	71.7	26,744	11,487	5,104	19	5	bottom trawl	1998–1999
Oregon & Washington[h]	143	71.7	10,253	10,108	2,246	22	5	shrimp trawl	1997–1999
Oregon & Washington	373	71.7	26,744	21,595	7,350	27	NA	bottom & shrimp trawl	1997–1999

[a]Data from Pilskaln et al. (1998) and NMFS data from 1991–1993.
[b]Data from Churchill (1989) and NMFS data from 1991–1993.
[c]Assumes 5 tows/fishing day, door spread of 150 ft., and 9 nm/tow.
[d]Assumes observed tows equal 0.75 actual tows, door spread of 600 ft., and 9 nm/tow; total tow distributed proportionally to observed tows.
[e]Assumes observed tows equal 0.75 actual tows, door spread of 550 ft., and 9 nm/tow; total tow distributed proportionally to observed tows.
[f]Assumes observed tows equal 0.40 actual tows, door spread of 400 ft., and 9 nm/tow; total tow distributed proportionally to observed tows.
[g]Assumes door spread of 300 ft. and 9 nm/tow.
[h]Assumes door spread of 150 ft. and 9 nm/tow.

NOTE: Relative intensity of trawling between regions based on assumptions regarding area swept, estimated total fishing area, and total number of trawl tows (used with permission from Natural Resource Consultants). NA indicates that data were not available.

of large scallops (Box 6.2). However, the only fishing effort data available for that region were collected before the areas were closed. In 1993, estimates of the percentage of statistical areas swept in New England waters indicated that the effort in some areas could have swept the grounds more than four times a year, and the effort in many statistical cells resulted in swept area estimates exceeding 100 percent of the block area.

Bottom trawling off the southeastern United States and in the Gulf of Mexico is for the most part concentrated in waters <20 m deep and close to shore. Trawling in the Pacific, North Pacific, and New England regions is directed primarily at groundfish and pandalid shrimps; bottom trawling in the Gulf of Mexico and off the southeastern states is directed primarily at various warmwater shrimp species, with some seasonal effort directed at crabs or scallops. The fishing gear used to

harvest shrimp is lighter than that used for most groundfish.

Data from the Gulf of Mexico show the average number of 24-hour fishing days in the region exceeded 200,000 annually from 1991 to 1999, averaging more than 4 million fishing hours per year. The data strongly suggest that the number of bottom trawl hauls per year (assuming five tows per 24-hour fishing day) in many of the statistical areas exceeded the bottom trawl effort off Alaskan coast, the contiguous Pacific states, or the New England region. The relatively large total area fished given in Table 4.1 for the Gulf of Mexico is partially an artifact of the division of the Gulf into 21 large fishing areas, some of which have very little effort. If the data were resolved into smaller statistical blocks, the total area fished would be much smaller and the percentage of the area swept per year would be

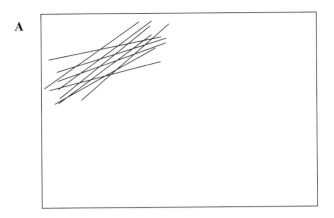

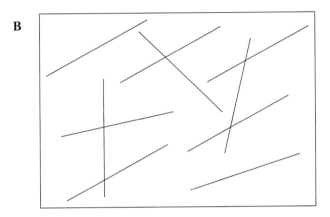

FIGURE 4.4 A) Trawl tracks concentrated in a favored fishing spot. B) Trawl tracks distributed throughout a reporting block.

higher in some blocks and lower in others. Local scientists have estimated that, from 1991 to 1999, some subareas in the Gulf of Mexico may have been swept more than 70 times a year.

In Alaska, the closure of several large areas has restricted trawling to a smaller portion of the total shelf and upper slope area. For example, in 1999–2000 trawling occurred in only 14 percent of the 702,898 km^2 area comprising the Bering Sea shelf and upper slope (National Marine Fisheries Service, 2001a). During the same period, 13 percent of the Gulf of Alaska waters (total area, 310,757 km^2) and 30 percent of the Aleutian Islands waters (total area, 59,124 km^2) were fished with bottom trawl or dredge gear. Currently, about 310,500 km^2 are closed year round to trawl gear in state and federal waters off Alaska (Figure 4.5).

The amount of bottom trawling declined significantly off Alaska, Oregon, Washington, and in the Gulf

of Mexico between the early 1990s and later in the decade (Figure B.38). Although the effort data have not been collected in recent years, this pattern is also likely for bottom trawling off the Atlantic coast as a result of more restrictive regulations during the late 1990s. Reductions in effort have followed declines in stock levels, closure of areas to trawling, and other management measures (Pacific Fishery Management Council, 2000).

Socioeconomic data are important in the evaluation of management alternatives that shift effort, gear, or fishing location. Selected fleet characteristics, including the number of vessels, crew sizes, and landed catch value for each fishery in each region of the United States, are summarized in Table 4.2. Trawl and dredge gears catch the majority of the food fish landed in the United States, and the fisheries employ thousands of people at sea and shoreside, including those who work in support services in coastal communities. The summary in Table 4.2 represents the best estimates based on existing data, but there is considerable variation in the availability and type of data collected for the different regions.

Because of the heterogeneity within and among fisheries, the costs and benefits of different management options will be borne differentially by fishery participants and their communities. For example, the economic consequences of gear modifications will be significantly different for small vessel, lower volume fisheries than for larger operations. Also, area closures have disparate effects on fishing communities and fishery participants, depending on the size and proximity of the closures and the characteristics of the fleet. Area closures can create safety problems as well as economic change, especially for smaller or less seaworthy vessels that must travel to more distant fishing grounds. It is important to acknowledge, however, that management alternatives also must be evaluated in terms of broader societal values and goals.

REGIONAL FISHERIES

New England: Maine to Connecticut

Fishery Descriptions: Groundfish Trawl, Northern Shrimp Trawl, Scallop Trawl, Scallop Dredge, and Clam Dredge

Trawling has been a dominant fishing method in the New England region for nearly a century. For many

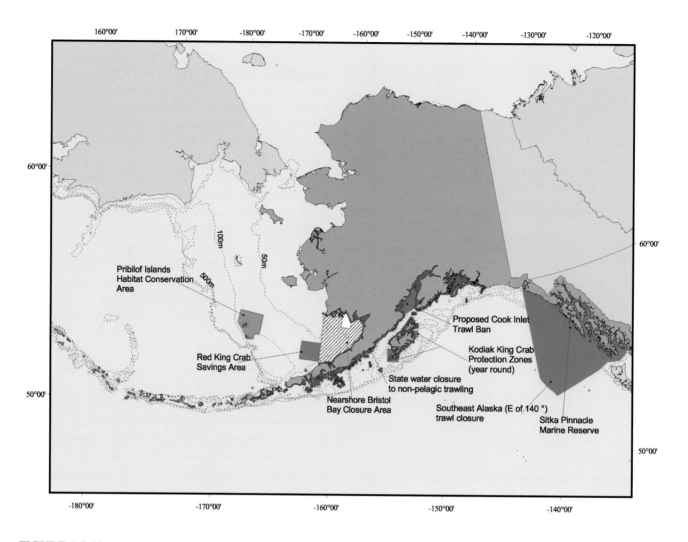

FIGURE 4.5 Year-round trawl closures areas in Alaska (Coon, 2001).

years, trawlers working the continental shelf from New York north to the Grand Banks made most of the nation's groundfish landings. The region is known for catches of cod, haddock, various flounders, and redfishes. The groundfish trawl fishery historically has been particularly important to New Bedford, Gloucester, and Chatham, Massachusetts; Portland, Maine; and Point Judith, Rhode Island (Aguirre International, 1996). These ports are home to highly specialized fleets. There were 748 federally permitted groundfish trawlers in the northeast to mid-Atlantic region (Maine to Cape Hatteras, North Carolina) in 1999. Trawling for northern shrimp also occurs seasonally in the Gulf of Maine, and there are 175 federally permitted vessels in that fishery.

The scallop fishery in 2000 was the most valuable in

the region. New Bedford is a major scallop port, producing 55 percent of scallops landed in the New England and mid-Atlantic regions (New England Fishery Management Council, 2000). There were 278 federally permitted scallop dredge vessels in the northeast region in 1999, and most of the permit holders have home ports in Massachusetts. Scallops are dredged and trawled; dredging accounts for approximately 80 percent of all landings from Connecticut to Virginia and is favored over trawling in New England. The fishery is seasonal: 90 percent of landings occur from June to November, regulated through a combination of trip limits and days-at-sea. Most scallopers participate in other fisheries, for monkfish, squid, dogfish, quahog, and summer flounder.

TABLE 4.2 Characterization of Trawl and Dredge Fisheries by Region

Region	Fishery	Catch		Crew Size	Number of Vessels	Vessel Size
		Pounds	Ex Vessel Value			
New England: Maine to Connecticut	Groundfish trawl	178,582,212	$132,230,808	2.1–6.6[a] (1)	748 (2)	45-100 ft. (2) or <5–>150 GRT (1)
	Northern shrimp trawl	3,483,277	$3,194,151	2.0–3.2[b] (1)	175[c] (1)	< 5–150 GRT (1)
	Scallop trawl	48,280	$79,649			
	Scallop dredge	23,543,729	$86,312,085	2–7 (1, 3)	278 (1)	5–>150 GRT (1)
	Clam dredge	20,193,963	$10,927,752	2.6–4.7[c] (1)	40 (1)	5–>150 GRT (1)
	Other mobile bottom gear	10,347,590	$4,622,331			
Mid Atlantic: New York to Virginia[d]	Finfish bottom trawl	72,755,587	$35,573,699	2.2–7.5[d] (1)		5–>150 GRT (1)
	Scallop trawl	837,156	$3,665,629			
	Scallop dredge	2,787,196	$11,627,885	2–7 (1, 3)	278 (1)	5–>150 GRT (1)
	Clam dredge[e]	66,476,284	$32,749,880	3.0–4.7[d] (1)	40 (1)	
	Other mobile bottom gear[f]	1,798,008	$2,492,792			
South Atlantic: North Carolina to Eastern Florida	Shrimp trawl	31,854,928	$77,802,322	3[g] / 3–4[h] (4)	2,400 (5) / 401 in 2000 and 133 in 1999 (4)	50–70 ft. (5) / 60–79 ft. (4)
	Calico scallop trawl	1,557,751	$1,724,468	2–6 (6)	26 fulltime and 40–50 parttime (6)	65–100 ft. (6)
	Finfish bottom trawl	10,601,327	$7,395,026			
Gulf of Mexico: Western Florida to Texas	Shrimp trawl	208,740,073	$439,654,593	1–3 (7)	3,598 (offshore)	25–85 ft.
	Calico scallop trawl	2,022,918	$1,721,757			56 ft. (mean)
	Finfish bottom trawl	8,271,722	$921,578			
Pacific: California, Oregon, and Washington	Shrimp trawl	29,979,426	$17,964,667	2–3 (OR) (8)	83[i] (CA) (9), 121 (OR), and 14 (WA) (10)	45–85 ft. (8, 11)
	Groundfish trawl	269,144,143	$58,877,066	2–3 (12)	397 (13) (in 1997)	63 ft. (mean) (in 1997)
North Pacific: Alaska	Finfish bottom trawl[j] (14)	875,266,000	$103,800,000	3–16	202[k]	56–155 ft.[l]
	Scallop dredge (15)	837,934	$2,982,760		10	90 ft. (mean)
	Shrimp beam and otter trawl[m] (10)	2,306,689	$859,133		21 beam and 5 otter (15)	

NOTE: The data have been aggregated from various sources and represent values for the 1999 fisheries unless otherwise indicated. Differences in data collection by local management agencies limit regional comparisons. For example, vessel size is measured by length in some areas and tonnage in others. These data are presented to provide the reader with an overview of the variety of trawl and dredge fisheries that operate in the United States. GRT means gross registered tons.

SOURCES: Landings by gear type for 1999 were taken from the NMFS commercial fisheries statistics database unless otherwise indicated (www.st.nmfs.gov/st1/commercial/landings/annual_landings.html). 1 = National Marine Fisheries Service, 2001b; 2 = Aguirre International, 1996; 3 = New England Fishery Management Council, 2000; 4 = South Atlantic Fishery Management Council, 2001; 5 = South Atlantic Fishery Management Council, 1996; 6 = South Atlantic Fishery Management Council, 2000; 7 = Thomas et al., 1995; 8 = Good et al., 1987; 9 = Thomson, 2001; 10 = Pacific States Marine Fisheries Commission, 1999; 11 = Leet et al., 1992; 12 = Pacific States Marine Fisheries Commission, 1998; 13 = Radtke and Davis, 2000; 14 = Hiatt et al., 2001; 15 = Alaska Department of Fish and Game, 2002.

continued

TABLE 4.2 Continued

[a]1999 average range for Tonnage Classes 1–4.
[b]1999 average range for Tonnage Classes 1–3.
[c]1999 average range for Tonnage Classes 2–4.
[d]Includes Chesapeake region.
[e]Landings data for all clam species.
[f]Landings data for oyster dredge, crab dredge, and beam trawls.
[g]Penaid shrimp only.
[h]Rock shrimp only.
[i]1995–1999 average.
[j]Includes sablefish, Pacific cod, flatfish, rockfish, and Atka mackerel.
[k]Groundfish trawlers (all species), excluding catcher/processors.
[l]Groundfish trawlers (all species), excluding catcher/processors. Mean for smallest to largest vessel length classes.
[m]Inshore, state managed fishery.

Effort Distribution and Intensity

Fishing effort in the New England region was analyzed based on port agent interviews and port landings, for the period 1991–1993. The effort data for various regions of the U.S. are reported in detail in Appendix B. During the early 1990s extensive areas of the continental shelf, extending east from 71°W longitude and including all of southern New England, Georges Bank, and the Gulf of Maine, were heavily trawled. In 1993, the 0.5° statistical blocks used for reporting effort were swept 2–4+ times by bottom trawls. On average, 56 percent of the region was swept more than once a year by trawl fishing gear. Analysis of similar data for the scallop dredge fishery indicates that effort is lower than in the trawl fishery, although scallop dredges cover an extensive area (Figure B.4). Effort data for this region in the late 1990s are unavailable, so it is not possible to analyze the effect of the area closures and effort reductions on the swept area estimates.

Mid-Atlantic: New York to Virginia

Fishery Descriptions: Mixed Finfish Trawl, Scallop Trawl, Scallop Dredge, Surf Clam Dredge, Winter Blue Crab Dredge (Estuarine), and Oyster and Clam Dredge (Estuarine)

Many areas of the mid-Atlantic region were fished before the 20th century. The region, often called the Middle Atlantic Bight, extends from Cape Hatteras to Cape Cod. Included are many large rivers and the Chesapeake Bay, the largest sheltered body of saltwater in the United States. The width of the continental shelf in this region varies from 30 miles in the south to 150 miles in the north. Most of the region's fisheries are relatively close inshore and include oyster, clam, crab, and menhaden. Trawling also occurs in this region for groundfish, scallops, and shrimp.

In mid-Atlantic estuarine waters, dredges are used to catch clams, oysters, conch, and crabs. Scallops and clams are the most economically important of the mobile bottom gear fisheries in the region; clams because of the high volume landed and scallops because of their high market value. Many scallopers in this region work part-time (New England Fishery Management Council, 2000). Unlike the New England fishery, in the mid-Atlantic region, scallop landings from trawling almost equal those from dredging. There are major scalloping areas offshore from the Delmarva Peninsula and the New York Bight that are second only to Georges Bank. Norfolk, Virginia accounts for 24 percent of landings from Connecticut to Virginia; Cape May, New Jersey accounts for 9 percent. Scallops fishing in the mid-Atlantic region is managed jointly with the New England Council, falling under the same days-at-sea and trip limit regulations. As in New England, the mid-Atlantic fishery is seasonal. About 80 percent of landings occur from May to August, so fishermen work in other fisheries to stay employed year round.

Trawling for groundfish is an important part of the local economy in many mid-Atlantic communities. In 1998, groundfish accounted for 86.8 percent of landed value in Freeport, New York; 74 percent in Shinnecock and Hampton Bays, New York; 59.9 percent in Montauk, New York; 50 percent in Belford, New Jersey; and 70 percent in Cape May, New Jersey (McCay and Cieri, 2000).

The surf clam and ocean quahog fisheries have operated under an individual transferable quota system since 1991. Since the 1970s, vessel sizes have increased steadily, and the number of vessels has decreased. There were 40 federally permitted vessels in the fishery in 1999. There are generally two ownership patterns: independent and processor owned (McCay and Creed, 1990). Independent vessels are based primarily in New Jersey. The largest processor fleets are in Maryland and Virginia.

Effort Distribution and Intensity

Fishing effort in the mid-Atlantic region was analyzed based on port agent interviews and port landings for 1991–1993. Details are reported in Appendix B. It appears that extensive areas of the mid-Atlantic continental shelf region were lightly trawled during this period as compared with the New England region. In 1985, 95 percent of the 0.5° statistical blocks were swept less than once a year by bottom trawls, and, on average, only 12 percent of the region's seabed was swept in 1985. Analysis of similar data for the scallop dredge fishery indicates that the effort also was considerably less intense than in New England; no statistical areas were swept more than once a year.

South Atlantic: North Carolina to Eastern Florida

Fishery Descriptions: Shrimp Trawl (Estuarine and Coastal), Crab and Fish Trawl (Estuarine), Rock Shrimp Trawl, Clam Dredge, and Calico Scallop Trawl

The South Atlantic region extends from North Carolina to the Florida Keys. Although this region is extensively trawled, almost all of the effort is carried out within shallow, inshore state waters and frequently within the sheltered waters of estuaries. For example, the North Carolina estuarine area is the largest of any in the Atlantic states, including more than 2.1 million acres of open water and almost 200,000 acres of wetlands. The protected inshore system is shallow, with an average depth of 3–4 m and otter trawls are used to catch shrimp and crab in North Carolina's estuaries. Between the Civil War and World War I, shrimp were caught with dip nets and seines. The first otter trawls were brought to the region in 1912. As early as 1925 more than 300 North Carolina fishermen were engaged

in the shrimp trawl fishery. It currently has more than 2000 participants. Close to 70 percent of the shrimp caught in or off North Carolina are taken in estuarine waters. Penaeid shrimp species are the region's most important mobile bottom gear fishery in terms of landings, employment, and ex-vessel value. South Carolina and Georgia fishermen primarily land white shrimp and have modified their trawls to fish off the bottom.

The Florida fishery is concentrated in the northeastern part of the state. It targets brown shrimp in the summer and white shrimp in the fall and winter. Most of the shrimping effort occurs in state waters (three miles on the east coast and nine miles on the west coast of Florida). Roller frame trawls are used to catch juvenile shrimp in seagrass beds to supply the recreational fishery with live bait. Rock shrimp and calico scallops also are trawled. Primary fishing grounds are hard sand, shell, and hash, but sometimes include corals. To reduce habitat damage to submerged corals, the South Atlantic Fisheries Management Council has recently expanded the habitat areas of particular concern for the Oculina Banks (South Atlantic Fishery Management Council, 2001) excluding trawling in this area. Additionally, the Council proposed a limited entry program for rock shrimp vessels. The rock shrimp season is from July through October, and most participants also fish for penaeid shrimp.

Historically, South Atlantic scallopers have used both dredges and trawls, although the fishery has been a trawl fishery exclusively since 1973 (South Atlantic Fishery Management Council, 2000). A directed fishery has been difficult to maintain because of fluctuating landings: some areas are productive one year and depleted the next. Consequently, with the exception of eastern Florida, scalloping historically has been a secondary fishery for bay shrimpers. Employment varies widely depending on stock abundance. In a good year, the income derived from scalloping can be as high as 75 percent of fishing-related income. There is currently one processor vessel. Virtually all landings occur in Florida, although fishermen and boats are concentrated in Florida and North Carolina.

Effort Distribution and Intensity

Quantifying the distribution and intensity of the mobile bottom contact fishing gear effort in the South Atlantic region is difficult because of the lack of comprehensive data (Appendix B). Most of the trawling and dredging occurs in state waters, if it occurs at all, and each state uses different methods of data collection, if and when data are available. Recently, this region has agreed to develop standardized reporting methods for trawling in both state and federal fisheries. In one 1995 data set from North Carolina, a localized, special inshore shrimp management area was documented to be swept more than 4 times by shrimp trawl gear. Most of the continental shelf off the North Carolina shore is hard rocky bottom inhabited by the snapper–grouper complex of fishes, but because North Carolina regulations prohibit trawl and dredge fishing vessels from possessing fish in this complex no effort is reported in this area.

Gulf of Mexico: Western Florida to Texas

Fishery Descriptions: Shrimp Trawl, Oyster Dredge (Estuarine), and Calico Scallop

Although there have been small-scale trawl fisheries targeting fish for food and industrial purposes in the Gulf, almost all bottom trawling currently is directed at various species of shrimp. Trawling is reported to have begun in the Gulf of Mexico around 1915, when shrimp seines were rapidly replaced with trawls (Shapiro, 1971). Following the introduction of trawling in the region, shrimp fisheries spread rapidly and by 1950 became Gulf-wide (Sheridan, 2001). Today, Texas, Louisiana, and Florida are home to largest shrimp fleets in the region (Adams, 1996). The Gulf shrimp fleet consists of relatively small boats, the average being 20 m. On average, captains have worked for more than 20 years in the fishery and almost half hire at least one relative as crew (Thomas et al., 1995). They work an average of nine months a year, making six trips a month, thereby specializing in shrimping (Thomas et al., 1995). In addition to the large directed shrimp fishery in the Gulf, there is also directed trawl and dredge fishing for scallops. All scallops are landed in Florida.

Effort Distribution and Intensity

The Gulf of Mexico trawling effort almost exclusively targets shrimp (Appendix B). The shrimp trawl fleet operates extensively throughout the Gulf, from nearshore to water depths of 90 m. The major areas of concentrated effort are middle depths off Florida, middle depths off Mississippi and Alabama, shallow depths off Louisiana (west Delta), and shallow and mid-depths off west Texas. During the 1990s, the shrimp fleet is reported to have averaged 204,000 24-hour fishing days, or about 4.9 million hours per year. For the period 1998–1999, this equates to the fished areas being swept more than 2.55 times annually, with 57 percent of the area being swept more than once annually. Comparing the periods 1991–1993 and 1998–1999, shrimp trawl effort has declined 41 percent, but it is reported that some localized areas in the Gulf of Mexico are swept 37–75 times per year.

Pacific: California, Oregon, and Washington

Fishery Descriptions: Finfish Trawl (Sablefish, Flatfish, Whiting, and Rockfish), Shrimp Trawl, and Sea Cucumber Trawl (California)

The trawl fisheries off California, Oregon, and Washington began early in the 20th century, generally in areas close to large population centers, such as Los Angeles, San Francisco, Portland, and Seattle. They have historically been conducted along the narrow continental shelf region that extends 20–30 miles off shore. The early fisheries were carried out in the shallow regions at depths of less than 100 m, but fishing extended into deeper waters as vessel size and horsepower increased, fishing methods and navigational aids improved, and the markets for a diversity of species expanded. Currently, most trawling occurs on the continental shelf at depths ranging from 15 m to more than 1000, with the vast majority between 15 m and 500 m.

Along the western U.S. coast, shoreside groundfish landings increased by 12 percent from 108,500 metric tons to 121,500 metric tons during 1983–1999, while shoreside ex-vessel revenues (in 1999 dollars) decreased by 47 percent from $100.2 million to $52.9 million (Pacific Fishery Management Council, 2000). These divergent trends were primarily the result of an increase in the landings of low-value Pacific whiting and a decrease in landings of high-value species, such as rockfish and flatfish. The abundance of many of the many groundfish stocks, with the exception of whiting, declined in the 1990s, and more restrictive fishery management measures have been imposed in an attempt to rebuild these groundfish stocks.

Groundfish yield both the highest value and the most landings for the mobile bottom gear fisheries. It operates under a limited-entry system and is managed with a combination of control strategies including trip limits and area closures. From the time that limited entry was instituted in 1994 through 2000, the trawl fleet declined by about 15 percent. The fleet is diverse in fishing patterns and in economic characteristics.

Oregon's Trawl Fisheries

In 1998, Oregon's commercial groundfish landings totaled 90,298 m, with an ex-vessel value of $23 million (Pacific States Marine Fishery Commission, unpublished report). Pacific whiting constituted 81 percent of these landings, with the remainder consisting mainly of flatfish, rockfish, and sablefish. The groundfish fishery is important to coastal communities, particularly Newport and Coos Bay.

The Oregon shrimp fishery is an important trawl fishery. Landings have cycled between 5 million and 10 million pounds (early 1970s, mid 1980s, late 1990s) and 45 million to 57 million pounds (late 1970s, late 1980s to early 1990s) since the late 1960s. Most harvests are taken off the central and northern Oregon coast, and shrimp landings are particularly important to the ports of Astoria and Tillamook.

Washington's Trawl Fisheries

In Washington State the seafood industry is important to several areas, mainly Ballard, Port Townsend, and Anacortes. There has been growth in the factory trawler sector that participates in the Alaska fishery as well as in whiting fisheries. Factory trawlers primarily home port in the Seattle area, and the smaller groundfish vessels come from other ports in northern Puget Sound. About 95 percent of the groundfish catch (by weight) for the Washington State fleet comes from Alaskan waters, mostly from the trawl fisheries (Natural Resource Consultants, 1999).

California's Trawl Fisheries

California's trawl fisheries began using the otter trawl in 1946 (Starr et al., 1998) and grew substantially after passage of the Magnuson-Stevens Act. Although groundfish landings have declined since 1981, they accounted for the state's highest average annual ex-vessel revenue by species group for the period 1995–1999 (Thomson, 2001). Together, shrimp and prawn ranked sixth and seventh in annual average revenue and weight landed for the same period. The minor sea cucumber fishery, located in southern California, accounts for a small proportion of California's fishery landings and revenues.

Patterns of fishery participation and dependence vary within and across California's three regions (Thomson, 2001). Thomson (2001) reports that an annual average of 29, 41, and 11 vessels participated solely in the groundfish trawl fishery in Northern, Central, and Southern California, respectively, between 1995 and 1999. Other fisheries in which groundfish trawl fishery participants take part include shrimp, prawn trawl, or crab pot fisheries in Northern and Central California. Trawl fisheries in Southern California are seldom pursued in combination with other gear types (Thomson, 2001). Mobility among regions, and in many cases among states, is critical to many California fishermen. For example, shrimp trawlers in Northern California earn more revenue from their out-of-state landings than their California landings (Thomson, 2001).

These characteristics of the California fisheries help explain the variability typical in fishing operations within and among fisheries. They also aid the consideration of opportunities, constraints, and potential responses to fishery management action. Those vessels that work a single fishery or gear, for example, could be less able to adapt to gear modification or reallocation of allowable catch to other gear types. Those who focus their effort in a single region are more vulnerable to local area or species closures than are those who fish in multiple regions. Operations that are more geographically diversified can intensify effort in one location if others are closed. Finally, although detailed information is lacking, bottom trawling supports not only the fishing operations—vessels and skippers, crews, and families—but also the people and businesses in coastal communities that receive and process their fish and provide support services (fuel, equipment, repair, and maintenance).

The groundfish fishery has recently been subject to severe regulatory restrictions that have widened the gap between harvest capacity and available harvest, thereby creating economic hardship for fishery participants, fishing households, and communities and greatly complicating monitoring and management of the fishery (Thomson, 2001).

Effort Distribution and Intensity

Information about the trawling effort off the coasts of California, Oregon, and Washington was provided by Natural Resources Consultants, the California Department of Fish and Game, the Oregon Department of Fish and Wildlife, and the Washington Department of Fish and Wildlife. Data from Oregon and Washington cover the periods 1991–1993 and 1998–1999; those from California cover 1994–1996. The major bottom trawled grounds in Washington were off Cape Flattery and to the southwest and from Westport to the Columbia River. The grounds off southern Washington were fished relatively heavily by vessels from Washington and Oregon ports. Off Oregon, the trawl fisheries are concentrated off major fishing ports such as Astoria, Newport, and Coos Bay. Important trawl grounds off California are, for the most part, found from Monterey north to the Oregon border, with relatively intense fishing between Santa Cruz and San Francisco and between Cape Mendocino and Crescent City. Two relatively heavily fished sites are also reported off Grover City and Ventura in southern California. State regulations prohibit trawling in state waters (3 nautical meters from shore) and on the continental shelf south of El Segundo to the Mexican Border; however, some trawling occurs on offshore banks in Southern California.

The intensity of trawling off the contiguous west coast states appears relatively similar for the three states, with slightly higher effort occurring off Oregon. (Note that the years for which data have been summarized for Washington, Oregon, and California differ.) The average number of tows per year off California from 1994 to 1996 was 15,535. It is estimated that there was a 60 percent reduction in effort between the early and late 1990s, because of declines in the abundance of target species and the subsequent Pacific Fishery Management Council reductions in quotas and fishing time. It is estimated that, during the mid 1990s, an average of 15 percent of the California shelf and slope area fished was swept more than once a year; 85 percent was swept less than once a year. Off the coasts of Oregon and Washington, during the late 1990s, 6 percent of the continental shelf and slope area fished was swept more than once a year; 94 percent was swept less than once a year (Figures B.35 and B.36).

A significant trawl effort also occurs off the three Pacific coast states for pink shrimp. The most intense fishing effort occurs off Oregon in the vicinity of Coos Bay, between Seaside and Garaibaldi, and off Pacific City. On average, about 10,000 shrimp tows per year were reported for areas off Oregon and Washington between 1997 and 1999.

North Pacific: Alaska

Fishery Descriptions: Finfish Bottom Trawl (Flounder, Yellow Fin Sole, Rock Sole, Rockfish, Atka Mackerel, and Cod), Scallop Fishery Dredge, and Shrimp Trawl (Otter and Beam)

The continental shelf and slope region off the coast of Alaskan coast constitutes one of the most extensive fishing grounds in the world. Bottom trawling in the Bering Sea began in 1929 with a Japanese exploratory operation; commercial operations began during the 1930s and early 1940s. Although these activities were terminated during World War II, they began again in the early 1950s (Alverson et al., 1964).

In the late 1950s, a large Soviet fleet entered the Bering Sea. Between 1960 and 1970, distant water bottom trawl fishing operations, conducted by several nations, intensely fished the Bering Sea and Gulf of Alaska. By the 1970s, foreign operations dominated bottom fishing on the continental shelf and slope in the Bering Sea and Gulf of Alaska. Domestic bottom trawling began with joint ventures in the Bering Sea in 1978 after passage of the Magnuson-Stevens Act in 1976. U.S. trawl activities in the Bering Sea and Gulf of Alaska grew rapidly during the 1980s, and, by the end of the decade, they had displaced foreign fisheries.

The most important trawl fisheries in Alaska in terms of total ex-vessel value, landings, and employment are for groundfish, which constituted 51 percent of the ex-vessel value for Alaskan fisheries in 2000. Bottom trawling provided 21 percent of the ex-vessel value—the major species are Pacific cod and flatfish (Hiatt et al., 2001). Pollock is the major groundfish species caught by trawl gear as measured by weight (71 percent in 2000) and value (65 percent in 2000), but in 1990 most of the catch (88 percent) was allocated to pelagic gear and by 1996 only 2 percent of the catch was caught by bottom trawl. In 1999, the use of bottom trawl gear was prohibited in the pollock fishery. Therefore, the pollock fishery is not included in this analysis.

In 1998, fishing cooperatives were introduced that use contractual agreements to manage catch and bycatch allocations between fishery sectors. The groundfish fleet is divided into catcher vessels and catcher processors, and in 1999, the catch is almost evenly divided between the two sectors (National Marine Fisheries Service, 2001c). There are 202 catcher vessels, most 18–37 m long (Hiatt and Terry, 1999). These vessels must carry the National Marine Fisheries Service observers 30 percent of the time. There are 40 catcher–processors ranging from 41 m to 130 m—most are longer than 54 m—that have 100 percent observer coverage (Hiatt and Terry, 1999). There are three "mother ships" that are longer than 133 m. Ownership of the groundfishing fleet is mixed; some vessels are owned by individuals, some by limited partnerships, and others by processing companies.

The state of Alaska manages the weathervane scallop fishery under the auspices of a federal fishery management plan that implements a license limitation program. The fishery is prosecuted by catcher–processor vessels that, with the exception of a few small vessels in Cook Inlet, are required to carry observers. The state of Alaska also manages shrimp fisheries throughout state and federal waters. Catches are taken mostly by otter trawl, beam trawl, and pot gear. The open-access shrimp fisheries currently have few participants because of low shrimp abundance, except in Southeast Alaska where more stable fisheries under a state-run limited entry program have sustained higher participation.

Effort Distribution and Intensity

The analysis of the intensity and distribution of fishing effort in the North Pacific was divided into three areas: the Bering Sea, the Gulf of Alaska, and the Aleutian Islands. The effort data include only trawl tows sampled by observers. Observer coverage includes 100 percent of the vessels longer than 38 m, and 30 percent of the 22–41 m vessels. Data were unavailable for vessels less than 22 m. Although the database is not complete, it is roughly representative of the spatial distribution and temporal changes in intensity.

Bottom trawling in the Bering Sea during the early 1990s was most intense on the slope and shelf area north of the Aleutian Islands and the Alaska peninsula in the vicinity of Unimak Island, east of the Pribilof

Islands, west of Bristol Bay, and off Cape Constantine (Appendix B). Large areas of the Bering Sea appear to have no trawling activity because of closed management areas, less productive fishing grounds, or unobserved tows. Both the spatial extent and the intensity of fishing effort decreased significantly in the 1990s. Between 1998 and 2000, 57 percent of the total area monitored was swept annually by bottom trawl gear; less than 2 percent of the area was swept more than once a year. The Gulf of Alaska experienced considerably less trawling activity than did the Bering Sea during the 1990s, and there were significant reductions in the geographic extent and the intensity of trawling in the Gulf of Alaska. The number of observed tows in the region was reduced by about half because of management area closures and because of general reductions in fishing effort associated with fisheries management. Bottom trawling off the Aleutian Islands extends from Unimak Island to 168°E longitude. The intensity of trawling was relatively light compared with the Bering Sea during the 1990s, and there was about a 40 percent reduction in observed effort during that decade.

CONCLUSION

Domestic trawl and dredge fisheries are conducted along most of the continental shelf and slope adjacent to the United States, although the level of fishing effort, and hence the amount of area affected, varies widely by region and by the spatial distribution of the fishing grounds. Groundfish trawls dominate the Alaska, Pacific coast, and New England fisheries; shrimp trawling is the major fishery in the Gulf of Mexico and in the coastal southeast Atlantic.

Over the past decade, management measures in some regions have closed areas to dragged gear to reduce gear conflicts, bycatch, or fishing mortality. Severe stock declines in some trawl and dredge fisheries have resulted in an overall reduction in bottom fishing effort. Although some areas of New England and the Gulf of Mexico are swept frequently, much of the continental shelf and slope area is trawled infrequently (less than one tow per year) if at all.

5

Analyzing the Risk to Seafloor Habitats

Seafloor habitats are subjected to a variety of fishery and nonfishery related stresses, and managers need a tool to assess their relative impact. Risk assessment is a flexible concept that has been applied to decision making in many fields, and several models that can be adapted to deal with ecological risk (e.g., National Research Council, 1983, 1993, 1996; Presidential/Congressional Commission on Risk Assessment and Risk Management, 1997). Risk assessment has been used in fisheries management (Fogarty et al., 1992, 1996; Smith et al., 1993) and can be considered as a part of an adaptive management framework (Walters, 1986) because it involves a risk or loss function. Adaptive management goes beyond risk assessment by explicitly including the feedback from policy decisions to the collection of new data and hypothesis testing (Walters, 1986).

FRAMEWORK FOR DECISIONMAKING

Ecological risk assessment is fundamentally a scientific undertaking that is the first step in the decision-making process. Risk management is informed both by ecological assessment and by the social, economic, and institutional features that constitute the human dimension of the issue. This should include characterization of the fishery participants and communities, fishery operations and practices, and associated institutions that govern fisheries. In addition to the ecological and social science assessments, risk management incorporates social values and legal mandates. Existing essential fish habitat (EFH) legislation is intended to protect the ecological function of habitat to support

fish production. In the longer term, the economic cost of failing to protect fish habitat would be forgone fish catches and related benefits to fishing communities as well as overall societal benefits. There are also economic costs related to other goods and services that marine ecosystems provide. At this stage, it is important to distinguish aesthetic and ethical values that do not have direct material value. In the context of seafloor habitats, there is a widespread drive to conserve marine biodiversity for benefits apart from concerns about fish population.

In the final stage of decisionmaking, a management strategy is chosen and regulations are enacted. The management options that can be used to address the effects of bottom trawling and dredging are described in detail in Chapter 6. At this stage, ecological and social sciences once again become important for developing new approaches to address the problem or to monitor the chosen measures. New research also might be required to clarify uncertainties, especially those about indirect effects on the resource species, on other parts of the ecosystem from which benefits could be derived, and on the response of the fisheries to new regulations as, for example, if disturbance of benthic spawning grounds by trawling led to a subsequent decline in fish recruitment. The slowed recovery of the exploited fish population could secondarily affect populations of prey species, and the continued decline in the catch could change the economic incentives of the fishery. As with aesthetic and ethical arguments, if the effects are considered large enough, new strategies for dealing with the issue and more research may be needed either to determine the best action or to measure success against objectives. Here again, the decision is outside the realm of science, because it reflects societal valuation, which is the domain of policymakers and stakeholders.

Ecological risk assessment is "the characterization of the adverse ecological effects of environmental exposures to hazards imposed by human activities" (National Research Council, 1993). This chapter describes two approaches to ecological risk assessment and discusses their utility for the management of benthic habitats. The exposure assessment model has been borrowed from the fields of human health and toxicology. It focuses on one risk at a time (National Research Council, 1983, 1993), it is quantitative, and it has been used in the policy arena to set standards and propose controls. The human health risk assessment framework was modified in 1993 for use in ecological risk assessment by the National Research Council's Committee on Risk Assessment Methodology. A modified model is described below.

The second method described here is comparative risk assessment. This method compares several types of risks and allows evaluation of the effects of a variety of stressors as opposed to a single stressor on seafloor habitat. Comparative assessments are used by policymakers to allocate resources and to set management priorities. In addition to data, they rely on expert judgment, scientific inference, and deliberation.

EXPOSURE ASSESSMENT MODEL

The exposure assessment model has three phases: research, risk assessment, and risk management (Figure 5.1). Policy mandates provide the regulatory framework of scientific research, monitoring, and validation, which provide important input at every stage. EFH provisions within the Sustainable Fisheries Act provide the context for risk management. The act requires the identification and minimization of threats to EFH. Evaluation of regulatory options to meet the

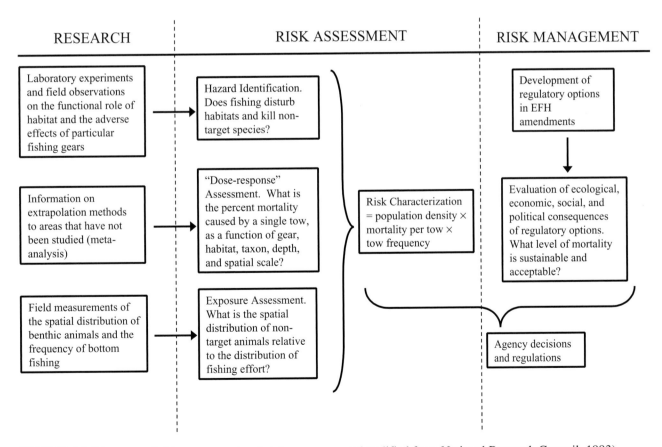

FIGURE 5.1 Elements of risk assessment and risk management (modified from National Research Council, 1993).

objectives requires not only ecological, but also economic, social, and political considerations. The product and outcomes of the risk assessment are agency decisions and regulations. In an adaptive framework there would be feedback from management decisions to further scientific research and assessment to aid in future decisionmaking (Walters, 1986).

Research

Scientific research provides the basic ecological information that feeds directly into the corresponding elements of risk assessment. Observational studies have documented associations between some fish species and structural components of their habitat. For example, laboratory experiments and field studies have documented the functional role of habitat structure in reducing the predation mortality of juvenile fish. Lindholm et al. (1999) reported that juvenile Atlantic cod in aquaria with simulated epifauna had lower predation mortality than they did in trials with a smooth sand substrate. Unfortunately, it is often impractical to study the functional role of fish habitat because of the difficulties associated with conducting experiments and making observations on the continental shelf and slope. Therefore it is necessary to extrapolate results from limited existing studies to areas that have not been studied (e.g., Lindholm et al., 2001).

One approach is to formulate testable hypotheses about how communities in different habitat types should respond to fishing (International Council for the Exploration of the Sea, 1996, 2000). If these hypotheses are supported, they can be used to extrapolate to unstudied areas. Meta-analysis—the summary of multiple, independent studies—can be used to identify the most important habitat variables and to construct quantitative models to predict fishing effects in unstudied areas (Chapter 3).

To estimate the mortality of nontarget species, it is necessary to know the spatial distribution and frequency of bottom fishing and the spatial distribution of the species of concern. These data must be coupled to assess the effects of trawling at the species and population level. Fishing effort data are being collected in observer programs and with vessel monitoring systems, with increasing spatial resolution (Rjinsdorp et al., 1998). The spatial resolution of benthic animals can be estimated with standardized research surveys, but those surveys generally use the same gear as used to capture the commercially targeted species. As a

result, there is often a paucity of data on nontarget benthic species.

Risk Assessment

Hazard identification is the determination of whether the ecological effects of the hazard (e.g., mobile bottom gear) are of sufficient concern to warrant further research or management. Hazards must have a detectable signature of ecological effects.

In the context of bottom-fishing effects, risk can be defined as the percentage mortality of a nontarget species (including structural epifauna that provide habitat for other organisms). This mortality has been measured in many experimental fishing studies and spatial comparisons of fished and unfished areas (see list of studies in Collie et al., 2000b). Most trawl impact studies have used standardized gear to simulate the disturbance caused by commercial trawling (e.g., Gordon et al., 1998); only a few studies have compared the relative effects of a particular gear (e.g., beam trawl) rigged differently (Bergman and van Santbrink, 2000).

Exposure response assessment is an important but poorly documented component of evaluating the effects of bottom trawling. Ideally, we wish to know the percentage mortality of a species caused by the single tow of a particular fishing gear as a function of habitat, depth, and so on. In practice, most trawl impact studies consist of a single treatment of one or more tows that might be more or less overlapping (Auster and Langton, 1999; Collie et al., 2000b). In a few instances controlled depletion studies have been conducted to measure the rate of disappearance of a species per tow (Poiner et al., 1998).

Exposure assessment involves calculating the overlap between the spatial distribution of fishing effort and the nontarget species. At this stage we can distinguish between *sensitivity* (an organism's innate ability to withstand physical damage) and *susceptibility* (the likelihood that an organism will be exposed to bottom fishing). Sensitivity is innate to the organism; susceptibility varies with intensity of impact. Thus, a sensitive organism in a nonimpacted area is not susceptible (International Council for the Exploration of the Sea, 1996).

Risk characterization involves the summary and synthesis of technical studies to describe the nature and severity of the risk. For bottom trawling, it means evaluating the mortality of each nontarget species at the population level. Risk characterization also addresses scientific uncertainty resulting from the application of experimental assessment studies to the real world (National Research Council, 1993). In some cases, this stage will be limited because of insufficient information to set quantitative reference levels.

Risk Management

It is at the risk management stage that ecological objectives are specified. One approach is to define reference points for mortality of nontarget species, similar to those developed for target fish species. There has been some progress based on the life-history characteristics of nontarget species (International Council for the Exploration of the Sea, 1998). Recognizing the practical impossibility of setting and monitoring compliance of reference points for every nontarget species in a community, the International Council for the Exploration of the Sea Working Group on Ecosystem Effects of Fishing Activities proposed the development of reference points for highly vulnerable indicator species (International Council for the Exploration of the Sea, 2000). That proposal assumes that the documented conservation of a set of highly vulnerable nontarget species gives a high probability of also conserving other, less vulnerable, nontarget species.

The exposure response method takes a species-by-species approach—each species would require its own reference. It would be possible to operate on habitat rather than by species. According to this approach, hazard identification could focus on the effects shown in a physical and biological index of habitat quality. Using this rationale, exposure assessment would tie risk (bottom trawling) to effect as measured on the index. Risk would be characterized by this surrogate for impact to the community in nontarget species. Decision-makers would use the index of habitat quality rather than population changes of nontarget species. The use of community level indices is appealing, but desirable or acceptable levels for these indices have not been developed.

Box 5.1 explains how such a calculation was made in the Dutch sector of the North Sea. Existing data could preclude a similar risk assessment for other areas, but it still could provide a useful framework for data collection. There are few, if any, examples in the United States where the annual mortality of nontarget species has been estimated. More information is required on extrapolating results to areas that have not

---

Box 5.1
Case Study: Mortality of Megafauna in the Dutch Sector of the North Sea

Although it was not presented as such (Bergman and van Santbrink, 2000), this example includes many of the elements of a risk analysis. The authors cite numerous studies of the effect of trawling in the Dutch sector, which collectively constitute the research stage of the risk assessment. The hazard identified is mortality of non-target invertebrates, either caught and discarded or killed as a result of direct contact with the gear or because of exposure to predators. Most direct mortality took place among animals in the trawl track and not in the bycatch. The exposure–response assessment consisted of estimating the direct mortality of invertebrates caused by the single passage of an otter or beam trawl. Direct mortality, expressed as a percentage of the initial density in the trawl track, ranged between 5 percent to 63 percent for some bivalve species. Exposure assessment was based on the spatial distribution of megafaunal invertebrates (>1 cm) and on the fishing frequency of trawl fleets in the statistical rectangles developed by the International Council for the Exploration of the Sea. In the risk characterization step, annual mortality due to fishing was calculated as the sum of specimens killed by a particular fishery, divided by the sum of actual numbers present in those rectangles. Annual mortality ranged from 5 percent to 39 percent, with half of the species having mortalities >20 percent. The authors' conclusions can be considered under the risk management heading. They recommend a framework in which ecosystem objectives are integrated with fisheries management. Possible regulatory actions are the same triad discussed in Chapter 6: reduction in fishing effort, more selective fishing gear, and closed areas. However, there is no explicit link between amount of annual mortality and regulatory action.

---

been studied. Also, more measurements are needed of the fine-scale distribution of the fishing effort. The potential hazards of bottom fishing have been identified, but more exposure–response assessment is needed. The regulatory options for risk management are well known. What remains unclear is an explicit relationship between the level of risk and the corresponding action. Thus more research is needed to link different levels of risk with particular mitigation strategies.

COMPARATIVE RISK ASSESSMENT

A risk assessment framework can be useful, even when the risks cannot be quantified with available data. Environmental policy is informed both by sound science and by societal values. Natural resources are a public trust, and environmental policy involves the disposition or distribution of those public resources. Thus environmental policymaking is both a social and a scientific venture. Comparative risk assessment is one way to use existing information effectively in managing trawl fisheries. Given finite resources and the amount of scientific uncertainty, it provides policymakers with a framework for setting priorities.

The comparative approach could be used to identify and assess the various sources of risk facing marine bottom habitat, including that posed by different types of gear, pollution, invasive species, and global warming. There are many methods used to structure comparative risk deliberations. They range from those that are data intensive, relying heavily on quantification, to those that rely more on deliberation among stake-

holders (Presidential/Congressional Commission on Risk Assessment and Risk Management, 1997).

A relational model that relies on available data, scientific inference, and deliberation is described below. An array of ecological risks is identified by a group of stakeholders drawn from the scientific community, agencies and local governments, the business community, and citizens. After the risks are identified, they are compared with one another against a set of criteria chosen by the stakeholder group. The criteria might include the scale of the disturbance, level of scientific uncertainty, immediacy of threat, irreversibility, and species affected. A scale from highest to lowest is constructed using the criteria. Criteria and measures of impact must be constructed to suit the nature of the risk (Table 5.1).

The method is described below, both with a hypothetical case study for application to bottom trawling using criteria and scales adapted from the Houston Foresight Project and with a real-world example used in the Alaska groundfish environmental impact statement.

TABLE 5.1 Matrix of Criteria and Measures of Impact

Criterion	Highest	High	Medium	Low	Least
Impact on size/ configuration	80 to 100 percent of area impacted	50 to 80 percent area impacted or highly fragmented	30 to 50 percent impacted or moderate fragmentation	10 to 30 percent impacted and some fragmentation	<10 percent impacted
Severity	Strong contribution to or cause of fully altered communities; degraded ecosystem	Creating built ecosystems; strong compromise to ecosystem integrity; or significant loss of community types	Compromise of community dynamics or loss of population resiliency	Modification of communities	Not a contributor to significant ecosystem or community modification
Uncertainty	Hard to imagine	Anecdotal	Based on accepted scientific model	Demonstrated for equivalent ecosystem	Demonstrated *in situ*
Immediacy	Already occurring	Starting now	Expected in near future	Expected sometime	May never happen
Irreversibility	Unrecoverable or >10 years	Recoverable 5 to 10 years	Recoverable 1 to 5 years	Recoverable approximately 1 year	Recoverable <1 year
Loss of human uses	50 to 100 percent capacity lost	10 to 50 percent lost	5 to 10 percent lost	1 to 5 percent lost	<1 percent lost

SOURCE: The example matrix above was adapted from the Houston Environmental Foresight Project (modified from Lester, 1995).

DESCRIPTION OF CRITERIA

Impact on size/configuration refers to the affected area. Is the risk confined to a small area, or does it extend over a large one? Is more than one type of habitat affected? Severity describes the nature of the effect and depends on the characteristics of both the risk and the habitat or species including: 1) substrate characteristics (sand, gravel, corals) and the seafloor's response to disturbance; 2) faunal and floral sensitivity to disturbance; 3) faunal and floral life history characteristics (long-lived attached versus mobile species); 4) frequency of fishing and gear characteristics; and 5) geographic range of benthic flora and fauna.

Uncertainty refers to the predictability of a particular result (habitat degradation) from a particular action (bottom trawling). Have the effects of a given risk been demonstrated in the ecosystem or area being assessed? If not, is it possible to make inferences from work conducted in other similar ecosystems?

Immediacy refers to the status of the risk. Is it occurring now? Is it likely to occur as a result of human population growth or other changes? For example, current freshwater inflows to a bay system could be adequate, but if projected increases in population for the next 20 years are expected to cause a concomitant increase in water use, the shortage of freshwater will become problematic.

Irreversibility refers to an ecosystem's or species' inability to recover from stress. Returning to the freshwater inflows example, oyster populations that are severely compromised one year because of salinity-induced disease can, with adequate fresh water, recover in a few years. Other species might take longer to recover from a drought or epidemic because of life history differences.

The human use criteria encompass consumptive and nonconsumptive uses of ecosystems and species. Such uses include fishing, scientific research, and recreation (diving, photography). The reduction of freshwater in a bay system will affect various commercial fisheries, but will not affect some kinds of recreation like boating.

Based on those criteria, each risk is assigned a score from highest to lowest. For example, using a comparative risk assessment, a manager could compare the risk to a particular area or ecosystem from trawling with the risks associated with other types of gear or with non-fishery related stresses like non-point source pollution. The example below assesses trawling and non-point source pollution and assumes that both currently occur on a fictional nearshore coral reef (Table 5.2).

TABLE 5.2 Ranking of Impacts on Nearshore Coral Reefs

Criterion	Trawling	Non-Point Source Pollution
Impacted Area	Low	Highest
Severity	Highest	Medium
Uncertainty	Least	Low
Immediacy	Highest	Highest
Irreversibility	Highest	High
Loss of human uses	Medium	High

1. Trawling
   - *Impact on size.* Localized trawling affects about 20 percent of the coral reefs.
   - *Severity.* However, each trawl inflicts severe damage by breaking corals, thus fully degrading that section of the reef and displacing its inhabitants.
   - *Uncertainty.* Biologic surveys in the area noted a rise in rubble areas with low biological diversity.
   - *Immediacy.* Trawling is already occurring, although it is seasonal.
   - *Irreversibility.* Because corals grow slowly, the recovery time will be long.
   - *Loss of human uses.* Human users associated with science and tourism avoid areas with damaged coral. Those uses represent 10 percent of the use of the reefs.
2. Non-point source pollution (soil and chemical runoff from nearby agricultural fields)
   - *Impact on size.* Runoff from spring rains increases the sediment and nutrient load over the entire reef.
   - *Severity.* Increased sediments and nutrient loads result in reduced light penetration; episodic hypoxia; a change in primary producers from benthic to planktonic organisms, increased incidence of toxic algal blooms, and sediment covering portions of the reefs.
   - *Uncertainty.* Research on the effects of non-point source pollution has not been conducted on this reef, but there is considerable scientific literature demonstrating the effects of runoff on other coral reefs. In situ, coral mortality is rising from unspecified causes.

- *Irreversibility.* Most of the reefs are experiencing stress, low growth, and some mortality. Because most corals are slow growing, recovery time is long.
- *Loss of human uses.* The reefs once supported a spear fishery and a large tourism industry. Tourism has declined by 30 percent; water visibility is low for part of the year and the recreational fishery has lower catch.

The stakeholder group uses the matrix to guide a discussion that considers scientific knowledge and data, scientific uncertainty, and social values. After each criterion is discussed and assessed, it is assigned a rank. After more deliberation, each risk receives an overall rank. The various risks are discussed, compared, and ranked by category. More than one risk can receive the same rank (e.g., highest or lowest). Some risks might be removed from consideration after being assigned a "least" ranking. There are no hard-and-fast rules for deciding the level of risk that requires action; that is decided by consensus, and is based upon the ranking.

A report on benthic EFH in Alaska (National Marine Fisheries Service, 2001b) can be considered an example of comparative risk assessment. Several fishery management alternatives were compared according to direct and indirect effects on benthic habitat. Numerical scores were assigned to the effects on habitat, with negative numbers representing greater risk, 0 the status quo management, and positive scores less risk (Table 5.3). Recognizing that ordinal scores are inherently subjective, the authors specified criteria to define moderate or marginal changes from the status quo. The management alternatives were then given scores according to the criteria (Table 5.4). The management alternatives considered for the Alaska groundfish fisheries can be summarized as follows:

1) continue status quo management policies,
2) emphasize protection of marine mammals and seabirds,
3) emphasize protection of target groundfish species,
4) emphasize protection of nontarget and forage species,
5) emphasize protection of habitat, and
6) increase socioeconomic benefits.

The table of scores illustrates the trade-offs in meeting fisheries management objectives (Table 5.4). Mea-

sures to protect marine mammals, seabirds (2), and target groundfish (3) would also have positive effects on habitat. Alternative 5 would shift fishing effort and its associated habitat effects from bottom trawl gear to fixed gear. Increasing socioeconomic benefits would increase the risk to benthic habitat because of greater bottom trawling effort. The ordinal scores can be used only to make ordinal comparisons. For example, a score of +2 means less risk than +1, but not necessarily twice as much habitat protection. It is therefore impossible to obtain overall comparisons between management alternatives by adding or averaging the scores (National Marine Fisheries Service, 2001b). This analysis was not meant to serve as a formal risk assessment, but it uses the same concepts to compare risks to benthic habitat associated with alternative fishery management policies.

SUMMARY

This chapter compares two methods used in ecological risk assessment—exposure response and comparative. The exposure response method is quantitative and is a modification of methods used to address toxicological risks. It quantitatively characterizes the response of a species or habitat to a particular stress.

The exposure response method has several limitations. Because it focuses on one risk at a time, other habitat stresses are not necessarily included as part of the context. Moreover, an accurate assessment depends on fairly complete scientific information about the relationship between the stressor (e.g., trawling), the biologic community, and the suite of processes necessary to a functioning habitat. Currently, there are gaps in our knowledge about the relationship between fish populations and their habitat and between specific stressors (e.g., gear) and habitat (Auster, 2001). These are identified more fully in Chapter 7. For now, an accurate exposure response assessment is not possible for all species and all habitats.

Comparative risk assessment is a qualitative method that compares different kinds of risk to each other and ranks them. It can be used when scientific knowledge is incomplete because it relies on a combination of available data, scientific inference, and public values. Additionally, managers get a more complete picture of all risks facing bottom habitats because each assessment addresses more than one risk. Managers also come to understand the public's view of what constitutes the greatest risks to bottom ecosystems. Com-

TABLE 5.3 Scoring System for Ranking the Effects of the Alternatives on Benthic Essential Fish Habitat

Issue	Direct Effects	Score −2	−1	0	1	2
Habitat Complexity	Removal/damage of HAPC biota by bottom trawl gear	Much more removal/damage to HAPC biota	Marginally more removal/damage to HAPC biota	Same level removal/damage as status quo	Marginally less removal/damage of HAPC biota	Much less removal/damage of HAPC biota
	Removal/damage of HAPC biota by fixed gear	Much more removal/damage to HAPC biota	Marginally more removal/damage to HAPC biota	Same level removal/damage as status quo	Marginally less removal/damage of HAPC biota	Much less removal/damage of HAPC biota
	Modification of nonliving substrates by bottom trawl gear	Amount of bottom trawl effort is more than 25 percent greater than status quo	Amount of bottom trawl effort 10 to 25 percent greater than status quo	Same (+/− 10 percent) amount of bottom trawl effort as status quo	Amount of bottom trawl effort 10 to 25 percent less than status quo	Amount of bottom trawl effort more than 25 percent less than status quo
	Modification of nonliving substrates by fixed gear	Amount of fixed gear effort is more than 25 percent greater than status quo	Amount of fixed gear effort 10 to 25 percent greater than status quo	Same (+/− 10 percent) amount of fixed gear effort as status quo	Amount of fixed gear effort 10 to 25 percent less than status quo	Amount of fixed gear effort more than 25 percent less than status quo

Issue	Indirect Effects	−2	−1	0	1	2
Minimization of Adverse Impacts	Benthic biodiversity	Area closed year-round to bottom trawl fishing more than 25 percent less than status quo	Area closed year-round to bottom trawl fishing 10 to 25 percent less than status quo	Same (+/− 10 percent) area closed year-round to bottom trawl fishing as status quo	Area closed year-round to bottom trawl fishing 10 to 25 percent more than status quo	Area closed year-round to bottom trawl fishing more than 25 percent greater than status quo

NOTE: HAPC = habitat area of particular concern.
SOURCE: National Marine Fisheries Service, 2001a.

parative risk assessment could be used to help identify the array of risks facing benthic habitat and to begin to prioritize these risks.

There are, however, limitations to this method. It is difficult to integrate various stressors into a single list because they affect ecosystems in different ways on different scales and they often interact. The scientific data that address different risks are often inconsistent or not comparable. Comparative risk assessment also lacks the scientific rigor of quantitative studies. The scoring of risks is inherently subjective; different groups will assign different scores. There is no consistent way to combine categorical scores to obtain an overall rank for each risk.

Risk assessment is valuable for linking scientific knowledge to management and public values, which should be used to identify and prioritize risks. There is, however, no best method that should be applied to all ecological problems. The method chosen depends on the quality and quantity of scientific data available and the policy and social contexts of the problems to be addressed.

TABLE 5.4. Scores for Each Alternative Reflecting Levels of Protection for Benthic Essential Fish Habitat Relative to Alternative 1

Issue	Effects	Alternatives 1	2.1	2.2	3	4.1	4.2	5	6.1	6.2
Direct Effects										
Habitat	Removal/damage of HAPC biota by bottom trawl gear	0	1	2	−1/1	0	0	2	−1/1	−2
Complexity	Removal/damage of HAPC biota by fixed gear	0	1	1/−1	1	0	1	−2	−1	−1
	Modification of nonliving substrates by bottom trawl gear	0	1	2	0	0	0	2	−1	−2
	Modification of nonliving substrates by fixed gear	0	2	2	1	0	2	−2	0	−2
Indirect Effects										
Minimization of Adverse Impacts	Biodiversity	0	0	0	2	0	0	2	0	0

NOTE: 0 = no change/difference from Alternative 1; −1/1 = marginal, or minor change from Alternative 1; and −2/2 = moderate or major change from Alternative 1. The index values contained in this table only contain ordinal information and can only be used to make ordinal comparisons. For example, an index value of 2 is better than a value of 1, but not necessarily twice as good. Therefore, it is not possible to obtain meaningful summary information by performing numerical operations (e.g., adding, subtracting, averaging) using two or more of the index values.

SOURCE: National Marine Fisheries Service, 2001a.

6

Management Options

The extent and effects of fishing on the seabed depend on gear design, access to fishing areas, and fishing effort. Three management tools for mitigating the effects of fishing on seafloor habitats correspond directly to those variables: modification of gear design or type, establishment of closed areas, and reductions of fishing effort (National Marine Fisheries Service, 1997). Management generally will warrant some combination of these measures. Because the social, economic, and regulatory context in which fishing occurs also influences the nature and extent of seafloor impacts, it is important to consider the opportunities for and constraints of particular management actions and their potential ecological as well as socioeconomic consequences.

Given the diversity of habitats, gear types, and interactions between them, and given the variety of social, economic, and regulatory contexts in which interactions occur, no single management solution will address all situations in the different regions and fisheries where use of mobile bottom gear affects the seafloor. Gear modifications often are more acceptable to a fishing community when they have fewer practical, social, or economic consequences. However, changes in gear tend only to diminish, not eliminate, seafloor impacts. Closed areas offer the advantage of eliminating the impacts caused by trawling and at particular sites, but closures also displace effort, potentially increasing fishing pressure elsewhere and causing economic and social problems in nearby coastal communities. Reduction of fishing effort can reduce the aggregated effects on seafloor habitat by decreasing the frequency and area of disturbance, but effort reduction

could be more difficult or costly from a human dimensions perspective. When the fishing capacity of the fleet is higher than necessary to harvest the allowable catch (overcapacity), as with the western groundfish fishery (Pacific Fishery Management Council, 2000), gear modification and area closures might be insufficient to mitigate seafloor habitat disturbance. Hence, in fisheries where overcapacity is a problem, effort reduction, in conjunction with area closures or gear restrictions, will be required both to sustain fisheries and to reduce seafloor impacts.

Each of these management tools is discussed in turn in this chapter. Some of the opportunities for and constraints to their use following from practical, social, and economic considerations are examined, and the potential ecological and human consequences of their implementation are discussed. Selected case studies are used to illustrate the application of management tools in various situations.

GEAR MODIFICATIONS

Gear modifications include changes in gear design, deployment, and type. Changes in gear design include alterations to existing gear, for example, by raising footropes on bottom trawls to reduce contact with the seafloor. Changes in gear deployment that could mitigate seafloor impacts include modifications in towing speed or duration. Changes in gear type include prohibition of some gears and reallocation to alternatives that cause less damage to seafloor habitats.

As described in Chapter 2, trawls and dredges have been developed and modified to enable fishing in less accessible, but often particularly valuable, habitats. In the North Irish Sea off the Isle of Man, for example, scallop vessels began fishing rougher areas of the seabed when fixed tooth-bar dredges were replaced with Newhaven spring-tooth dredges, introduced in 1972, coupled with a reduction in dredge size and an increase in the number of dredges fished in a spread (Brand, 2000; Mason, 1983). In the Northwest Atlantic and elsewhere, the development of rockhopper gear with 24 inch rollers allowed trawl vessels to drag through rough bottom types. Before those innovations, the costs associated with the higher frequency of gear loss or damage prevented most fishermen from fishing in these areas and generally limited the scope (if not the magnitude) of seafloor impacts.

Most gear modifications have been motivated by economics. Especially in areas where stocks have de-

clined or where demand has surpassed local supply, the drive to catch more fish has created an incentive to modify gear to fish more efficiently or to access previously unfished sites. In other cases, economic, regulatory, and other incentives have encouraged gear modifications to promote conservation and increase marketable catch. In some instances, the prospect of limited fishing opportunities because of unacceptable bycatch rates has prompted technological innovations toward gear that generates less bycatch and reduces seafloor contact. The Alaskan pollock fishery (Box 6.1) provides a case study of an incentive-based gear innovation that was driven by a need to reduce bycatch.

Gear modifications or innovations come from within and outside the fishing industry. In the case of the Alaska pollock fishery, the fishermen were given the incentive to reduce bycatch, but they also were given

---

Box 6.1
Case Study: Gear Modifications in the Alaskan Pollock Fishery

The walleye pollock (*Theragra chalcogramma*) fishery of the eastern Bering Sea is one of the largest in the world. In 2000, 1.1 million metric tons of pollock was captured. Pollock occur on the sea bottom and midwater up to the surface, and most catches are taken at 50–300 m. The fishery is managed with total allowable catch (TAC) for the target species, constrained by bycatch limits for several pelagic and demersal species.

In 1990, concerns about bycatch and seafloor habitats affected by this large fishery led the North Pacific Fishery Management Council to apportion 88 percent of TAC to the pelagic trawl fishery and 12 percent to the nonpelagic trawl fishery (North Pacific Fishery Management Council, 1999). For practical purposes, nonpelagic trawl gear is defined as trawl gear that results in the vessel having 20 or more crabs (*Chionecetes bairdi*, *C. opilio*, and *Paralithodes camstschaticus*) larger than 1.5 inches carapace width on board at any time. Crabs were chosen as the standard because they live only on the seabed and they provide proof that the trawl has been in contact with the bottom.

By the mid-1990s, most vessels participating in the pollock fishery had voluntarily switched to pelagic trawls. Prohibited species bycatch limits provided the incentives: If the limits were exceeded as recorded by onboard observers, premature fishery closures would take effect before the pollock TAC was taken. Even though nonpelagic trawls accounted for only 2 percent of the pollock catch in 1996, they were nearly one-third of the halibut bycatch and about one-half of the crab bycatch. One year later, out of continuing concerns about bycatch and the effects of trawl gear on the seafloor, the Alaska Marine Conservation Council proposed that the North Pacific Fishery Management Council ban all bottom trawling for pollock. In response, the North Pacific Fishery Management Council prepared an amendment to the Bering Sea and Aleutians Islands groundfish fishery management plan (North Pacific Fishery Management Council, 1999). In November 1999, with broad industry and public support, the North Pacific Fishery Management Council banned bottom trawl gear use in the Bering Sea pollock fishery. The fishery now attains TAC specifications with modest bycatch rates.

Although this gear was modified to reduce bycatch, it is postulated to have had the secondary effect of diminishing the impact on seafloor habitat. However, these trawls may be frequently fished in contact with the seafloor, especially in shallow water (<50 fathoms). To confirm that this gear has reduced seafloor impacts, the extent of bottom contact and disturbance should be quantified. If the trawls never touch the bottom, the pelagic trawl definition could be set at zero crab tolerance. Because typical pelagic trawls have large mesh webbing in the lower section of the net and are affixed to chain footropes, bycatch enumerated by onboard observers might substantially underestimate the number of demersal fish and invertebrates that are affected because they fall through the large mesh panels instead of being captured by this gear.

the latitude to develop technology and practices to achieve that goal. Their direct involvement in the process facilitated practical and acceptable changes.

In contrast, specific gear modifications to reduce bycatch and exclude turtles from trawls in the Gulf of Mexico and the Southeast were imposed by state and federal agencies. Initially, these measures were strongly resisted by the industry, in part because of their impracticality, but also because they told fishermen how to fish, and thereby dismissed a key area of fishermen's knowledge and expertise.

Another fundamental constraint to gear modifications could be a lack of awareness or public recognition of a particular kind of gear, and the potential benefits of modifying its design or deployment to mitigate those effects. Visual presentations of how gear alters the seafloor can be instructive, both in making fishermen and others aware of the problem and in stimulating discussion about potential gear modifications.

The development and testing of fishing gear technology is expensive and time-consuming. In some cases, it is difficult for fishermen to experiment with new gear designs, especially if they participate in highly competitive, open-access fisheries. However, other opportunities for innovation can be created. One way would be to establish a limited experimental fishery in which gear could be tested without loss of fishing opportunity. This approach was used to develop and conduct at-sea trials of the raised footrope trawl in the New England silver hake fishery. Another mechanism is to study gear–habitat interactions, funded by landings taxes or flat assessments (like the California Salmon Stamp), or by tax credits given to industry for sponsored gear research. Finally, academic, government, and commercial research facilities for testing and computer modeling of new gear designs can provide further opportunities for the development of modifications. Collaborations among gear technologists, fishermen, and net manufacturers have been successful in addressing concerns about gear selectivity.

Gear modification will not be an appropriate solution to bottom habitat damage in all cases, however, either because it fails to diminish damage or because it is physically, socially, or economically impractical. Some species (e.g., scallops, flatfish) can only be captured by mobile bottom-contact gear. Where gear modifications are technologically feasible, social and economic considerations can prevent their adoption. As with other management measures, gear modifications entail several costs. Those include not only the

financial costs of modifying the gear, but also those associated with learning how to use the gear effectively and with the possibility of reduced catch efficiency. In addition, although some modifications may improve the quality of the catch, others result in reductions in either quality or quantity that are unacceptable to fishermen, fish buyers, and consumers. Some otter trawls, for example, are designed to cause a cloud of sediment that herds fish into the trawl path (Smolowitz, 1998). Gear modifications that reduce habitat disturbance are likely to reduce catch rates, and therefore would be unacceptable to most fishery participants. Requirements to use less efficient gear to protect habitat could lead to more intensive fishing effort, thereby offsetting the benefits.

CLOSED AREAS

Closed areas encompass regions of the seafloor where mobile bottom-contact gear is not allowed, either permanently or temporarily. The recent National Research Council report, *Marine Protected Areas: Tools for Sustaining Ocean Ecosystems* (National Research Council, 2001), defines several types of closed areas, differentiated by their goals and the degree of protection sought. The report defines marine protected areas (MPAs) as discrete geographic areas that have been designated to conserve and enhance marine resources through an integrated plan that includes restrictions on some activities. Marine reserves are MPAs in which some or all biological resources are protected from removal or disturbance. Marine reserves include fishery reserves, which preclude fishing for some or all species to protect critical habitat, rebuild stocks, and protect against overfishing, and ecological reserves, zones that protect all living marine resources from removal or disturbance other than for research purposes to evaluate reserve effectiveness.

As spatially based management tools, MPAs are consistent with the concepts of essential fish habitat and habitat areas of particular concern (HAPC). As stated in the interim final rule, HAPC can be designated based on one or more of the following criteria: 1) the importance of the ecological function provided by the habitat, 2) the extent to which the habitat is sensitive to human-induced environmental degradation, 3) whether and to what extent development activities are or will stress the habitat type, and 4) the rarity of the habitat type. The recent Tortugas Ecological

Reserve and other MPAs along the coast of the United States were established in response to many of these concerns.

Interest has been growing in the potential role of MPAs in fishery and broader marine resource management, and there has been a proliferation of efforts to establish them within and outside the United States. To date, most closed areas were implemented to reduce fishing mortality, at least within the reserve, rather than to protect habitat, per se. Enhanced public awareness of the adverse effects of mobile bottom gear on seafloor habitats, however, has increased interest in, and the use of, marine reserves to protect bottom habitats from these effects (National Research Council, 2001). Direct evidence of the structure and complexity of some habitats can enhance recognition of their vulnerability to mobile bottom fishing gear and engender support for area closures to protect them. For instance, photographic documentation of red tree corals (*Primnoa willeyi* and *P. resedaeformis*) and associated long-lived *Sebastes* species and other fishes led to broad public support for the creation of the Sitka Pinnacles Marine Reserve in the eastern Gulf of Alaska (O'Connell et al., 1998). Only pelagic troll gear for salmon is allowed in this reserve.

However, MPAs raise important social and economic issues that warrant careful consideration. One potential negative effect of closed areas is crowding in the areas that remain open to fishing. Although overcrowding might not be problematic from a habitat perspective if vessels are displaced into less-sensitive habitats, it can still have negative social and economic consequences. Displacement of fishing effort can lead to incursions into other fishermen's or other resource users' (e.g., recreational users') territory, creating social conflict both on the fishing grounds and at the docks or market. Informal territorial use rights in fisheries (LeVieil, 1987) and other spacing conventions have been widely documented, especially in fixed gear fisheries (e.g., Maine lobster fishery, Acheson, 1975, 1988; Alaska golden king crab [*Lithodes aequispinus*] fishery, Herrmann et al., 1998), but also in such mobile gear fisheries as the cod trawl fisheries off Newfoundland (Durrenberger and Palsson, 1987). Furthermore, closures could have economic costs both for those who have been displaced and for those who work the areas that remain open. Moreover, if there is a large displacement of effort, intensified fishing in open areas can result in ecological damage, including overfishing of other stocks. These concerns suggest that area

closures should be combined with effort reduction, gear modification, or both, to reduce potential ecological disturbance, although the social and economic consequences of the combined measures would need to be assessed. Other social consequences of closed areas include loss of access and increased costs and risk, if fishermen must travel to more distant or more dangerous fishing grounds.

Enforcement is an extremely important consideration determining the efficacy of closed areas. Closures are much more likely to be successful when they have the support of the fishing industry; the cooperation of affected users is essential to ensuring compliance. Adequate funding for enforcement operations also is important. New technologies, such as satellite transponders and satellite-mounted synthetic aperture radar (for viewing vessels through clouds and at night), can be effective enforcement tools.

Georges Bank provides a good example of the use of closed areas for fishery management (Box 6.2). Initially there was great opposition, but over time, this management tool has become accepted by most fishermen as benefits have accrued from improved stocks and higher catch rates for some species. But even as fishermen hail the economic benefits of better catches outside the closed areas, some are pressing for a partial reopening to obtain even higher catches.

Rotational area closures, a variant of marine reserves, have been implemented to afford some protection to seafloor habitats while not permanently closing access. Rotational closures also can be more consistent with some fishing patterns. In some fisheries (e.g., scallop dredge fisheries), fishermen are known to "give areas a rest," rotating their effort among locations to adapt to spatial and temporal variations in resources. The New England Fishery Management Council is currently considering an amendment that would include rotational area closures in the management plan for the scallop fishery (New England Fishery Management Council, 2002).

Because rotational closures allow periodic fishing, they are inappropriate for highly structured seafloor habitats with long-lived attached species. But they could be viable for more energetic, sandy habitats inhabited by short-lived species. In applying rotational closures, schedules for closing and opening areas should be tied to recovery time. In shallow areas with frequent storms, the recovery time might be very short, as the fauna and flora have adapted to natural disturbance. Rotational closures, based on regular monitoring, fit within an adaptive management framework.

Box 6.2
Case Study: Closed Areas on Georges Bank

In response to the collapse of the principal groundfish species—cod, haddock, and yellowtail flounder—the Secretary of Commerce took emergency action in December 1994 by initiating year-round closure of two areas on Georges Bank and one in southern New England (Figures B.4 and B.5). These areas, totaling 17,000 km², were closed to all bottom-fishing gear capable of catching groundfish, and they have remained closed except for partial and temporary openings for scallop dredging in 1999 and 2000 (Murawski et al., 2000). During the same period, fishing effort was reduced by half for most of the mobilegear fleets, and complementary regulations were implemented on the Canadian side of Georges Bank.

It appears that the implementation of those management measures has allowed scallop and some groundfish stocks to rebuild substantially. Standardized surveys conducted by the National Marine Fisheries Service show much higher densities of groundfish and scallops inside the closed areas. The degree of protection afforded is related to the proportion of the stock contained in the areas and the fraction of the year they reside in the area. The closed areas have been most successful in the conservation of the more sedentary demersal fishes and sea scallops. Haddock and yellowtail flounder have recovered to an abundance last observed in the 1970s; between 1994 and 1998, scallop biomass increased 14-fold in the closed areas (Murawski et al., 2000). The recovery of cod has been slower because of the lack of strong recruitment.

The area closures, combined with effort reductions in the fishery, have reduced fishing mortality on the principal groundfish stocks, and have protected the seafloor habitat from the physical effects of bottom fishing. Current studies are comparing the benthic communities inside and outside of the closed areas. Particularly in the northern part of Closed Area II, there has been a rapid increase in epifauna on gravel sediments. In 1998, the New England Fishery Management Council designated part of the closed area as an HAPC on the basis of the occurrence of juvenile groundfish on gravel–cobble sediment.

The success of these management measures is largely attributable to the closure of areas with the highest groundfish and scallop catch rates. Simultaneous effort reduction measures (fewer days fished) helped reduce the consequences of displaced effort. Some boats switched to different fisheries, although they bear higher costs because of increased effort in other areas or fisheries. The current fishing effort is concentrated around the edges of the closed areas, which suggests that they are acting as sources for the surrounding areas. In the future, the boundaries of the closed areas could be refined to enhance larval production and protect nursery areas, spawning concentrations, and migration corridors (Murawski et al., 2000).

EFFORT REDUCTION

Fishery managers often strive to reduce effort as a way to eliminate biological or economic overfishing. Recruitment overfishing occurs when spawning stock biomass is reduced so much that future recruitment is compromised. Growth overfishing occurs when fish are caught before they grow large enough to achieve maximum yield per recruit, but without decline in recruitment levels (Gulland, 1983). Economic overfishing occurs when excess fishing effort causes a fishery to produce no positive economic rent, that is, when the total costs of extraction equal or exceed the revenue provided by the fishery (Clark, 1976).

In addition to reducing the ecological and socio-economic impacts of overfishing, effort reduction can lessen the effects of trawling and dredging on the seafloor. Effort can be reduced through seasonal closures, license limitations, quotas, vessel buyback programs,

or trip limits. Often they are used in combination, as when limited entry is combined with a fishery quota to guard against excessive effort by those who remain in the fishery. The establishment of some form of rights-based fishery management program (e.g., individual fishing quotas) is one approach for meaningful and permanent reduction of fishing effort (National Research Council, 1999).

Figure 6.1 illustrates the relationship between fishing effort and seafloor habitat disturbance from mobile bottom-contact gear. As effort increases, so does habitat damage until all epibenthic structure and associated biota have been removed. At that point, the curve levels off because maximum habitat damage has occurred. At high effort levels, reductions will decrease damage marginally at first, and benefits will increase as effort declines further. The amount of fishing effort at which maximum habitat damage occurs (Figure 6.1, B) depends

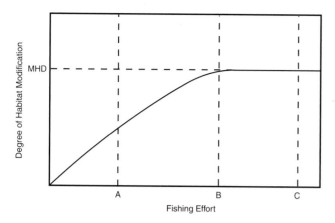

FIGURE 6.1 Schematic of the relationship between fishing effort and habitat damage. When effort corresponds to A, a change in effort will change the degree of habitat damage. At B, damage is at its maximum and increased effort does not increase damage, bur reduced effort decreases damage. At C, small adjustments in effort neither increase nor decrease habitat damage. MHD refers to maximum habitat damage.

on habitat type and the amount of natural disturbance. If the effort level is high (Figure 6.1, C), it could be impractical to decrease effort to the extent that habitat will begin to recover.

Effort reductions can decrease effects in sensitive areas if they result in a smaller area being swept by fishing gear. As noted in Chapter 3, the need or desire to increase catches has led to increases in effort and expansion into new, and sometimes more sensitive, habitats. Effort reduction could slow or arrest this process; decrease the incentive to develop new, more intrusive gear; and limit or reduce the spatial extent of trawling and dredging and hence their disturbance of seafloor habitat. As stocks recover, the fishing fleet is likely to concentrate in high-density areas, thus decreasing the total area of seafloor disturbed by fishing.

There are important practical, social, and economic considerations that must attend effort reduction strategies. Reductions in effort usually result in immediate short-term losses of income, employment, and lifestyle for at least some fishery participants, even if they hold the promise of longer term benefits to fishermen, resources, and habitat. Those consequences raise questions about equity, manifested in the allocation of costs and benefits that society and policymakers will need to address. Moreover, if initial measures, such as trip

limits, fail to conserve fish stocks, social and economic consequences could be exacerbated by later, more substantial effort reductions. Those could include limited-entry programs that substantially cut the number of fishery participants and the aggregate fishing capacity. On the West Coast, years of increasingly strict effort controls in the face of declining groundfish fish populations were unsuccessful in matching capacity with the level of the resource. After the declaration of the West Coast groundfish disaster in February 1999, the Pacific Fishery Management Council's Scientific and Statistical Committee determined that only 27–41 percent of the trawl fleet's current capacity was needed to catch its allocation (Pacific Fishery Management Council, 2000). The Council is now considering how to eliminate excess capacity in the fishery to reduce effort, with the expected benefit that less effort also will diminish seafloor disturbance.

Practical, social, and economic considerations warrant attention not only in the context of the fishery in question, but also in the broader context of regional fisheries. Participants displaced from a trawl fishery might move into other fisheries, possibly causing further ecological, economic, and social problems. Ideally, decisions to reduce fishing effort, as with all proposed major changes in fishery management systems, should be informed by analyses of the full suite of short- and long-term benefits and costs.

A case study of the Browns Bank scallop fishery (Box 6.3) provides an example of a technological solution to achieve effort reduction that also reduces the effects of fishing gear on the seafloor. The same quota was caught and less area of bottom was affected with the multibeam technology (1999) than without it (1998). However, there are caveats to the universality of this approach. First, success hinges on the implementation of TAC that is set as a sustainable fraction of scallop biomass. In the absence of TAC, fishing fleets could use the technology to deplete fishery resources more efficiently. Second, the technology successfully identifies the densest aggregations of scallops, but it is not known whether those high catch rates can be sustained over the long-term. Third, such industry–government collaborations are exemplary, but in such cases the resultant data could be proprietary, leading to policy or legal issues concerning data access by other potential users. Finally, scallops predominate on gravel rather than sand substrate, and more concern exists about scallop dredge effects on hard than on soft bottoms. It has yet to be demonstrated or quantified

---

Box 6.3
Case Study: Fishing Effort Controls in the Browns Bank Scallop Fishery

The Canadian scallop dredge fishery on Browns Bank on the western Scotian Shelf northeast of Georges Bank provides an example of a technological approach to reducing the total amount of seafloor swept by mobile bottom-contact gear through de facto effort controls (Kostylev et al., in press; Manson and Todd, 2000). The fishery is important locally, accounting for approximately one-third of the region's shellfish catch. Since the 1970s, fishing has been prosecuted on different portions of the bank, with inconsistent success. The stock abundance is estimated from assessment surveys, stratified largely by the distribution of the commercial fishing effort. Recently, this fishery has been managed on the basis of an enterprise allocation system, in which each of seven companies receives a share of the annual TAC (Department of Fisheries and Oceans Canada, 2000).

A recent collaboration among the Department of Fisheries and Oceans, the Geological Survey of Canada Atlantic, and the fishing industry is exploring the application of geosciences to the fishery. The project's objectives include documenting the relationships among scallops and substrate, optimizing fishing practices, and adopting sustainable fishery management through increased knowledge. The project has entailed intensive data collection from multibeam bathymetry, high-resolution seismic reflection, sidescan sonar, extensive bottom sampling, video, and photographic surveys. Recently, a scallop-catch sampling program was added to the research. The research demonstrates that scallops are strongly associated with gravel lag deposits, which the multibeam data easily distinguishes from sandy bottom. There is a highly significant relationship between backscatter intensity and scallop survey catches that could be incorporated into improved stock assessments.

Although the industry's prime motivation initially was to improve efficiency, other benefits have accrued, as evidenced by the following tabular comparison of fishery attributes from 1998, when multibeam maps were not used, and from 1999, when multibeam maps were applied during the fishery.

	1998	1999
Scallop quota, kg	13,640	13,640
Time on bottom, hr	162	43
Distance towed, km	1176	311
Time lost, hr	15	0
Value of lost gear	$10,000	0
Fuel used, L	72,697	17,545

Application of the technology resulted in a 73 percent reduction in both the duration of bottom contact time and in the area of habitat affected, a 75 percent reduction in fuel use, and an elimination of gear loss and lost fishing time. The implication is that habitat disturbance can be substantially reduced if information about the relationship between the substrate type and scallop abundance is used to target fishing effort to the most productive scallop grounds.

---

that overall ecological damage is reduced when effort is reduced but concentrated on gravel bottoms. The amount of damage caused by mobile bottom-contact gear depends on the frequency of repeated trawling (or dredging) and the recovery time of affected fauna. Whether it is better to spread the effort or concentrate it into a few, heavily affected areas is an important, but complex, question. Notwithstanding those qualifiers, the Browns Bank scallop habitat project is an excellent example of a collaborative and technological approach to meet management goals for seafloor habitats.

CONCLUSION

Three fishery management tools can be used to mitigate the effects of trawls and dredges on seafloor habitats, fishing effort reduction, modification of gear design or gear type, and area closures. Three fishery management tools, fishing effort reductions, modifications of gear design or gear type, and establishment of areas closed to fishing, are used to mitigate the effects of mobile bottom-contact gear on seafloor habitats.

Effort reduction is the cornerstone of managing the effects of fishing, including, but not limited to, the effects on habitat. However, effort reduction alone is insufficient to address all circumstances in which fishing gear disturbs bottom habitat. The success of fishing effort reduction depends on the resilience and recovery potential of the habitat. Each of the other management tools generally requires effort reduction to achieve maximum benefit.

Gear modifications will be most useful for finfish species that can be caught with gear that does not rely on disturbing the bottom to catch the fish. This could include shifts to a different gear type, such as long lines or fish traps, but the social and economic consequences of such reallocation must be recognized and addressed. Also, the overall ecological benefits of using another, and often less efficient, type of gear, can be reduced if there is a subsequent increase in fishing effort or if there is significantly higher bycatch with the alternative gear.

Closed areas are necessary to protect a range of representative habitats. Closures are particularly useful for protecting areas with emergent epifauna (e.g., corals, bryozoans, hydroids, sponges) that are vulner-able to even low levels of fishing effort. As evidenced by the case of Georges Bank (Box 6.2), damage to emergent epifauna is recoverable after areas are closed. In general, area closures will need to be paired with effort reductions to offset the effects of displaced effort in the open fishing grounds.

It is unlikely that any one measure can resolve all seafloor habitat issues. Rather, some combination of options will often be most effective. The choice, utility, and limitations of a particular combination of the three measures to control fishing effects on seafloor habitats in a specific situation depends on the current regulatory setting, social and economic characteristics of the fishery and its participants, available habitat types, and the specific fishery management goals and objectives. Ideally, the choice of the particular mix of the three tools for any one case should be informed by analyses of the full suite of benefits and costs over a reasonable period. As demonstrated by the case studies, creative solutions can be found to lessen the effects of fishing on seafloor habitats while maintaining viable, long-term commercial fisheries. In fact, the two are inextricably connected.

7

Findings and Recommendations

The most challenging aspect of evaluating the effects of trawling and dredging on seafloor habitats is translating observed effects from experimental studies to the scale of actual fishing effort in the various fisheries around the United States. Studies that examine changes in seafloor structure and biological communities after disturbance by various types of mobile fishing gear have yielded consistent patterns of acute effects that can be categorized by gear type, habitat characteristics, composition of the benthic community, and frequency of disturbance. To convert those results into an assessment of ecosystem-level effects on seafloor habitats requires analysis of the frequency of bottom trawling and dredging and fine-scale mapping of this effort relative to the geography of seafloor habitats in the fishing grounds. Further research will be necessary to fully understand the effects of chronic disturbance by mobile bottom gear and to more accurately assess the effects of habitat disturbance on the productivity of commercial and recreational fisheries.

The acute, gear-specific effects of trawling and dredging on various types of habitat are well documented (Chapter 3). Many studies indicate that stable communities of low mobility, long-lived species are more vulnerable to acute and chronic physical disturbance than are communities of short-lived species in changeable environments. Habitat complexity is reduced by towed bottom gear that removes or damages biological and physical structures. The extent of the initial effects and the rate of recovery depend on the stability of the habitat. The more stable biogenic, gravel, and mud habitats experience the greatest changes and have the slowest recovery rates. In con-trast, less consolidated coarse sediments in areas of high natural disturbance show fewer initial effects. Because these habitats tend to be populated by opportunistic species that recolonize more rapidly, recovery also is faster.

Data on the geographic distribution and frequency of trawling and dredging are limited in spatial resolution, and there is considerable regional variation in reporting methods and in records for recent years (Chapter 4). However, data collected in the early 1990s indicate that the most intensive effort (Table 4.1) was in the fishing grounds of the Gulf of Mexico and New England regions. Bottom trawling in the mid-Atlantic, Pacific, and North Pacific regions was relatively light, with less than 1 tow per year in many reporting areas. Even in heavily trawled regions, effort was not evenly distributed; thus, some areas were trawled several times per year while others were trawled infrequently if at all. Throughout the 1990s and into 2001 there were significant reductions in the intensity and spatial extent of bottom trawling (Figure B.38). These decreases resulted from reductions in fishing effort, area closures, and gear restrictions instituted by managers in response to problems with declining fish stocks, bycatch, or interactions with endangered species.

The largest information gap is in the spatial distribution of different habitat types in trawled or dredged areas. For most areas only coarse maps are available on habitat distribution. This mismatch in the spatial scales of experimental results, habitat maps, and trawl effort reporting data makes it difficult to accurately assess effects of trawling and dredging on marine ecosystems. Nonetheless, there is enough information currently available to support efforts to improve the management of the effects of fishing gear on seafloor habitats. Specific recommendations for making the best use of current information and suggestions for research are provided below.

RECOMMENDATIONS

The following recommendations fall into three categories: 1) interpretation and use of existing data; 2) integration of management options; and 3) policy issues raised by existing legislation. Recommendations for research appear at the end of this section.

Interpretation and Use of Existing Data

Recommendation

Fishery managers should evaluate the effects of trawling based on the known responses of specific habitat types and species to disturbance by different fishing gears and intensity of fishing effort, even when region-specific studies are unavailable. The direct responses of benthic communities to trawling and dredging (Chapter 3) are consistent with ecological models of how biological communities and ecosystems respond to acute and chronic physical disturbance. Although area-specific studies on the effect of trawling and dredging gear will allow more targeted management approaches, adequate information is available to address fishing effects on seafloor habitat. Predictions developed from common trends observed in comparable habitats will provide reasonable estimates of fishing effects to serve as the basis for management. Estimates should be revised as more site-specific information becomes available. An adaptive management strategy could be used both to reduce the effects in the short-term and to provide additional information for improving management in the long-term.

Recommendation

The National Marine Fisheries Service and its partner agencies should integrate existing data on seabed characteristics, fishing effort, and catch statistics to provide geographic databases for major fishing grounds. The potential consequences of fishing can be most efficiently assessed by the simultaneous and consistent presentation of all available data on the characteristics of the seabed and fishing effort. Although some data exist on habitat characteristics and on the location and intensity of fishing (Chapter 4), the

available data have been prepared by different agencies, in different formats, at variable levels of resolution, and are collected in separate archives. Integration of these databases into a single, geographic information system will assist managers in evaluating regional needs for habitat conservation.

Integration of Management Options

Recommendation

Management of the effects of trawling and dredging should be tailored to the specific requirements of the habitat and the fishery through a balanced combination of the following management tools.

- *Fishing effort reductions.* Effort reduction is the cornerstone of managing the ecological effects of fishing, including, but not limited to, effects on habitat. Other management tools (gear restrictions or modifications and closed areas) may also require effort reduction to achieve maximum benefit. However, effort reduction alone might not be sufficient to reduce effects in highly structured habitats where there is low potential for recovery.
- *Modifications of gear design or restrictions in gear type.* Disturbance depends on the extent of contact of the gear with the seafloor; gear designs that minimize bottom contact can reduce habitat disturbance. In addition, shifts to a different gear type or operational mode can be considered, but the social, economic, and ecological consequences of gear reallocation should be recognized and addressed.
- *Establishment of areas closed to fishing.* Closed areas effectively protect biogenic habitats (e.g., corals, bryozoans, hydroids, sponges, seagrass beds) that are damaged by even minimal fishing.

The appropriate combination of management approaches will depend on the characteristics of the ecosystem and the fishery—habitat type, resident seafloor species, frequency and distribution of fishing, gear type and usage, and the socioeconomics of the fishery. Each characteristic should be evaluated during development of a mitigation strategy.

Recommendation

The regional fishery management councils should use comparative risk assessment to identify and evaluate risks to seafloor habitats and to rank

management actions within the context of current statutes and regulations. Risk assessment provides a scientifically informed approach to clarifying environmental policy issues by elucidating the environmental consequences of particular policy choices (Chapter 5). Comparative risk assessment can be used when there is incomplete scientific information because it relies on a combination of available data, scientific inference, and public values. Mobile bottom gear is only one of many factors contributing to the degradation of benthic habitats; the comparative approach provides a method for simultaneous consideration of a wide range of risks, including pollution, drilling, and natural disturbance. Comparative risk assessment enables stakeholder involvement in the decision-making process and improves the sharing of information among different groups to aid in the development of solutions that have broad societal support.

Policy Issues Raised by Existing Legislation

Recommendation

Guidelines for designating essential fish habitat (EFH) and habitat areas of particular concern (HAPC) should be established based on standardized, ecological criteria. The EFH concept recognizes that management of exploited fish populations requires addressing effects on other parts of the ecosystem upon which fish depend. Its effective use rests upon a clear understanding of the population biology and the spatial distribution of each managed species. The current designation of EFH does not require the use of consistent criteria with respect to the assignment of habitat to each life stage of species covered by fishery management plans. Instead, the regional councils develop the criteria, often based on data availability. Typically, EFH designations are too extensive to form a practical basis for managing fisheries (Chapter 1). Although this approach could assist in mitigating some habitat threats, it provides little guidance for evaluating the effects of trawling and dredging. For example, in some management plans habitat is identified based on the frequency with which fish are found in a particular area. Although this method is based on sound ecological principles, it is important to refine the use of frequency distributions to identify the habitats that support the main fraction of the population rather than to simply document where the fish are found.

The term HAPC should be clearly and narrowly defined with establishment of specific guidelines for regulating fishing activities. The effectiveness of the designations should be reviewed periodically. HAPC forms a subset of EFH based on the ecological value of the area, its susceptibility to perturbation, and whether it is rare or currently stressed (National Marine Fisheries Service, 1997). However, current policy does not require additional protection for HAPC. Because of the demonstrated importance of HAPC in the life cycles of exploited fish populations, HAPC sites should receive priority in fishery management plans.

Recommendation

A national habitat classification system should be developed to support EFH and HAPC designations. A classification system with common habitat designations will improve efforts to protect, inventory, and construct regional or national maps of habitats of importance or concern. Standardizing classifications would facilitate tracking changes over time and provide the basis for developing functional links between the underlying mechanisms that structure the ecosystem and the biological systems that support fisheries production. A habitat classification system would assist in ranking the relative importance of different habitats for fisheries and for biodiversity, estimating the vulnerability of the habitat to disturbance, and facilitating the application of research conducted in one region to the management of habitats in other regions.

FUTURE RESEARCH

In the course of this study, many gaps were identified in the current understanding of the effects of fishing on the seafloor. The following recommendations are intended to direct research towards filling these gaps. They have been organized into three primary areas of research—gear impacts and modification, habitat evaluation, and management—with some overlap between categories.

Gear Impacts and Modification

Further research on gear effects will be required to develop a predictive capability to link gear type and effort to bottom disturbance, fish production, and recovery times in particular habitats. Active engagement of resource users in the research will help ensure that mitigation strategies are practical, enforceable, and acceptable to the fishing community. This could be accomplished through cooperative research programs involving fishermen and scientists. Topics for future research include the following:

- identification of the forces produced by fishing gear on the seafloor and the threshold forces that injure and dislodge a range of benthic organisms;
- use of various empirical approaches, such as sidescan sonar, to assess the spatial extent and overlap of trawl and dredge effects in conjunction with higher resolution effort reporting data;
- development of fishing gear to reduce damage to habitat and to meet other conservation goals such as bycatch reduction and maintenance of biological communities; and
- investigation of why some areas appear to continue to produce fish despite chronic disturbance by fishing gear.

Habitat Evaluation

Habitat disturbance has been studied mainly at small spatial scales with short-term observations of acute disturbance. Development of a landscape-scale perspective of the effects of trawling and dredging on the seabed will require a long-term commitment to higher resolution mapping of the continental shelf and slopes. Because most studies have focused on animal communities, more studies are needed on ecosystem processes (e.g., productivity, nutrient regeneration). Topics for future research include the following:

- the rates and magnitude of sediment resuspension, nutrient regeneration, and responses of the plankton community in relation to gear-induced disturbance;
- the dose-response relationship as a function of gear, recovery time, and habitat type to evaluate effects of repeated disturbance by fishing gear;
- recovery dynamics, with consideration given to estimating large-scale effects at current fishing intensities;
- acute and chronic effects of trawling in deeper water (>100 m);
- recovery rates in stable and structurally complex habitats;

- relative magnitude of different sources of bottom habitat disturbance;
- long-term trend data for benthic production versus fisheries production; and
- the effects of habitat fragmentation on total production.

Management

Constructive interactions among stakeholders and policymakers can be facilitated through user group funding of research and through collaborative research that involves scientists and fishermen. Increased participation also will support cooperative development of alternative gears and practices. Comparative risk assessment can be used to identify priorities for acquiring quantitative data to improve risk analysis. Monitoring and evaluation of the consequences of existing management measures (e.g., gear restrictions, area closures, effort reductions) should be used to support the development of new management plans, especially in understudied regions. Aggregation and analysis of existing information on habitats, fishing effort, and efficacy of various management measures will help the regional fishery management councils meet their mandate to protect EFH. Topics for future research include the following:

- develop testable hypotheses of how biological communities in different habitat types respond to fishing;
- establish baselines for characteristic habitats and regions to document the effects of various fishery practices;
- design quantitative models to predict fishing effects in areas that have not been studied;
- validate the use of frequency-dependent distribution approaches for designating EFH and HAPC through analysis of community structure and life history parameters; and
- collect and analyze data on the social and economic characteristics of trawl, dredge, and non-mobile gear fisheries to assess the tradeoffs among various management alternatives.

References

Acheson, J.M. 1975. The lobster fiefs: Economic and ecological effects of territoriality in the Maine lobster industry. *Human Ecology* 3(3):183-207.

Acheson, J.M. 1988. Patterns of gear changes in the Maine fishing industry: Some implications for management. *MAST/Maritime Anthropological Studies* 1(1):49-65

Adams, C. 1996. An overview of the commercial and recreational fisheries industries within the Gulf of Mexico. *The Southern Business and Economic Journal* 11:246-260.

Aguirre International. 1996. *An Appraisal of the Social and Cultural Aspects of the Multispecies Groundfish Fishery in New England and the Mid-Atlantic Regions.* Report to NOAA for contract # 50-DGNF-5-00008.

Allee, R.J., M. Dethier, D. Brown, L. Deegan, R.G. Ford, T.F. Hourigan, J. Maragos, C. Schoch, K. Sealey, R. Twilley, M.P. Weinstein, and M. Yoklavich. 2000. *Marine and Estuarine Ecosystem and Habitat Classification.* NOAA Technical Memorandum NMFS-F/SPO-43.

Alverson, D.L., P.T. Pruter, and L.L. Ronholt. 1964. *A Study of Demersal Fishes and Fisheries of the Northeastern Pacific Ocean.* Institute of Fisheries, University of British Columbia.

Auster, P.J. 1998. A conceptual model of the impacts of fishing gear on the integrity of fish habitats. *Conservation Biology* 12:1-6.

Auster, P.J. 2001. Defining thresholds for precautionary habitat management actions in a fisheries context. *North American Journal of Fisheries Management* 21:1-19.

Auster, P.J., R.J. Malatesta, and S.C. LaRosa. 1995. Patterns of microhabitat utilization by mobile megafauna on the southern New England (USA) continental shelf and slope. *Marine Ecology Progress Series* 127:77-85.

Auster, P.J., R.J. Malatesta, R.W. Langton, L. Watling, P.C. Valentine, C.L.S. Donaldson, E.W. Langton, A.N. Shepard, and I.G. Babb. 1996. The impacts of mobile fishing gear on the sea floor habitats in the Gulf of Maine (Northwest Atlantic): Implications for conservation of fish populations. *Reviews in Fisheries Science* 4:185-200.

Auster, P.J., C. Michalopoulos, P.C. Valentine, and R.J. Malatesta. 1998. Delineating and monitoring habitat management units in a temperate deep-water marine protected area. In *Linking Protected Areas with Working Landscapes, Conserving Biodiversity,* Munro, N.W. and J.H.M Willison (eds.). Science Management of Protected Areas Association, Wolfville, Nova Scotia.

Auster, P.J. and R.W. Langton. 1999. The effects of fishing on fish habitat. In *Fish Habitat: Essential Fish Habitat and Rehabilitation,* Benaka, L. (ed.). American Fisheries Society, Bethesda, MD.

Barnette, M.C. 1999. *Gulf of Mexico Fishing Gear and Their Potential Impacts on Essential Fish Habitat.* NOAA Technical Memorandum NMFS-SEFSC-432.

Barnette, M.C. 2001. *A Review of Fishing Gear Utilized within the Southeast Region and Their Potential Impact on Essential Fish Habitat.* NOAA Technical Memorandum NMFS-SEFCS-449.

Bergman, M.J.N. and J.W. van Santbrink. 2000. Mortality in megafaunal benthic populations caused by trawl fisheries on the Dutch continental shelf in the North Sea in 1994. *ICES Journal of Marine Science* 57(5):1321-1331.

Berkeley, S.A., D.W. Pybas, and W.L. Campos. 1985. *Bait Shrimp Fishery of Biscayne Bay.* Florida Sea Grant college Program Technical paper No. 40.

Bradstock, M. and D. Gordon. 1983. Coral-like bryozoan growths in Tasman Bay, and their protection to conserve commercial fish stocks. *New Zealand Journal of Marine and Freshwater Research* 17:159-163.

Brand, A.R. 2000. North Irish sea scallop fisheries: Effects of 60 years dredging on scallop populations and the environment. In *Alaska Department of Fish and Game and University of Alaska Fairbanks: A Workshop Examining Potential Fishing Effects on Population Dynamics and Benthic Community Structure of Scallops with Emphasis on the Weathervane Scallop Patinopecten Caurinus in Alaskan Waters.* Alaska Department of Fish and Game, Division of Commercial Fisheries, Special Publication 14, Juneau.

Breitburg, D.L. 1998. Scaling eutrophication effects between species and ecosystems: the importance of variation and similarity among species with similar functional roles. *Australian Journal of Ecology* 23:280-286.

Bridger, J.P. 1970. *Some Effects of the Passage of a Trawl over the Seabed.* Gear and Behavior Committee, International Council for the Exploration of the Sea.

Brylinski, M., J. Gibson, and D.C. Gordon, Jr. 1994. Impacts of flounder trawls on the intertidal habitat and community of the Minas Basin, Bay of Fundy. *Canadian Journal of Fisheries and Aquatic Science* 51:650-651.

Butman, B., P.C. Valentine, W.W. Danforth, L. Hayes, and L.A. Serrett. In press. *Shaded Relief, Backscatter Intensity, and Sea Floor Topography of Massachusetts Bay and the Stellwagen Bank Region, Offshore of Boston, Massachusetts.* U.S. Geological Survey Geologic Investigation Map I-2734.

Caddy, J.F. 1973. Underwater observations on tracks of dredges and trawls and some effects of dredging on a scallop ground. *Journal of Fisheries Research Board of Canada* 30:173-180.

Caddy, J.F. 1993. Toward a comparative evaluation of human impacts on fishery ecosystems of enclosed and semi-enclosed seas. *Review of Fisheries Science* 1:57-95.

Cahoon, L.B., R.L. Redman, and C.R. Tronzo. 1990. Benthic microalgal biomass in sediments of Onslow Bay, North Carolina. *Estuarine Coastal and Shelf Science* 31:805-816.

Cahoon, L.B. and J.E. Cooke. 1992. Benthic microalgal production in Onslow Bay, North Carolina. *Marine Ecology Progress Series* 84:185-196.

Cahoon, L.B., G.R. Beretich, Jr., C.J. Thomas, and A.M. McDonald. 1993. Benthic microalgal production at Stellwagen Bank, Massachusetts Bay, USA. *Marine Ecology Progress Series* 102:179-185.

Canadian Fishery Consultants Limited. 1994. *Fishing Gear and Harvesting Technology Assessment.* CFCL, Halifax, Nova Scotia.

Churchill, J.H. 1989. The effect of commercial trawling on sediment resuspension and transport over the Middle Atlantic Bight continental shelf. *Continental Shelf Research* 9(9):841-864.

Churchill, J.H. 2001. *Sediment Resuspension by Bottom Fishing Gear,* [Online]. Available: http://www.fishingnj.org/artchurchill.htm [2002, February 26].

Clark, C.W. 1976. *Mathematical Bioeconomics: The Optimal Management of Renewable Resources.* John Wiley and Sons, New York, 352pp.

Collie, J.S. and P.D. Spencer. 1994. Modeling predator-prey dynamics in a fluctuating environment. *Canadian Journal of Fisheries and Aquatic Sciences* 51:2665-2672.

Collie, J.S., G.A. Escanero, and P.C. Valentine. 1997. Effects of bottom fishing on the benthic megafauna of George's Bank. *Marine Ecology Progress Series* 155:159-172.

Collie, J.S., S.J. Hall, M.J. Kaiser, and I.R. Poiner. 2000a. A quantitative analysis of fishing impacts on shelf-sea benthos. *Journal of Animal Ecology* 69:785-798.

Collie, J.S., G.A. Escanero, and P.C. Valentine. 2000b. Photographic evaluation of the impacts of bottom trawling on benthic epifauna. *ICES Journal of Marine Science* 57:987-1001.

Connor, D.W., K. Hiscock, R.L. Foster-Smith, and R. Covey. 1997. A classification system for benthic marine biotopes. In: *Proceedings of the 28th European Marine Biology Symposium.* Crete, September 1993. Olsen and Olsen, Fredensberg.

Coon, C. 2001. *Managing Effects of Bottom Trawling on Marine Habitat in Alaska.* Presentation to the Committee on Ecosystem Effects of Fishing: Phase 1—Effects of Bottom Trawling on Seafloor Habitat, Anchorage, AK, June 2001.

Dayton, P.K. 1979. Observations of growth, dispersal, and population dynamics of some sponges in McMurdo Sound, Antarctica. In *Biologie des Spongiaries,* Levey, C. and N. Boury-Esnault (eds.).

de Groot, S.J. 1984. The impact of bottom trawling on benthic fauna of the North Sea. *Ocean Management* 9:177-190.

DeAlteris, J., L. Skrobe, and C. Lipsky. 1999. The significance of seabed disturbance by mobile fishing gear relative to natural processes: A case study in Narragansett Bay, Rhode Island. *American Fisheries Society Symposium* 22:224-237.

Deegan, L.A. and R.N. Buchsbaum. 2002. The effect of habitat loss and degradation on fisheries. In *The Decline of Fisheries Resources in New England: Evaluating the Impact of Overfishing, Contamination, and Habitat Degradation,* Buchsbaum, R.N., W.E. Robinson, and J. Pederson (eds.). Massachusetts Bays Program. MIT Sea Grant Press, Cambridge, MA.

Department of Fisheries and Oceans. 2000. *2000 Scotia-Fundy Offshore Scallop Integrated Fisheries Management Plan: Maritimes Region.* DFO, Canada.

Duarte C. 1995. Submerged aquatic vegetation in relation to different nutrient regimes. *Ophelia* 41:87-112.

Durrenberger, E.P. and G. Palsson. 1987. Ownership at sea: Fishing territories and access to sea resources. *American Ethnologist* 14(3):508-522.

Dyekjaer, S.M., J.K. Jensen, and E. Hoffmann. 1995. *Mussel Dredging and Effects on the Marine Environment.* ICES C.M.1995/E:13.

Emeis, K.C., J.R. Benoit, L. Deegan, A.J. Gilbert, V. Lee, J.M. Glade, M. Meybeck, S.B. Olsen, and B. von Bodungen. 2001. Group 4: Unifying concepts for integrated coastal management. In *Science and Integrated Coastal Management,* von Bodungen, B. and R.K. Turner (eds.). Dahlem University Press, Berlin, 378pp.

Engel, J. and R. Kvitek. 1998. Effects of otter trawling on a benthic community in Monterey Bay National Marine Sanctuary. *Conservation Biology* 12:1204-1214.

Fogarty, M.J., A.A. Rosenberg, and M.P. Sissenwine. 1992. Fisheries risk assessment: Sources of uncertainty. *Environmental Science Technology* 26(3):440-447.

Fogarty, M.J., R.K. Mayo, L. O'Brien, F.M. Serchuk, and A.A. Rosenberg. 1996. Assessing uncertainty and risk in exploited marine populations. *Reliability Engineering and System Safety* 54:183-195.

Fogarty, M.J. and S.A. Murawski. 1998. Large scale disturbance and the structure of marine systems: Fishery impacts on Georges Bank. *Ecological Applications* 8(1 Supplement):S6-S22.

Fonseca, M.S., G.W. Tanyer, A.J. Chester, and C. Foltz. 1984. Impact of scallop harvesting on eelgrass meadows: implications for management. *North American Journal of Fisheries Management* 4:286-293.

Freese L., P.J. Auster, J. Heifetz, and B.L. Wing. 1999. Effects of trawling on seafloor habitat and associated invertebrate taxa in the Gulf of Alaska. *Marine Ecology Progress Series* 182:119-126.

Frid, C.L.J. and R.A. Clark. 2000. Long-term changes in North Sea benthos: Discerning the role of fisheries. In *Effects of Fishing on Non-Target Species and Habitats: Biological, Conservation, and Socio-Economic Issues,* Kaiser, M.J. and S.J. de Groot (eds.). Blackwell Science, Oxford.

Futch, C.R. and D.S. Beaumariage. 1965. *A Report on the Bait Shrimp Fishery of Lee County, Florida.* Florida Board of Conservation Marine Laboratory Maritime Base, Baybora Harbor, St. Petersburg, Florida.

Good, J. W., R.G. Hildreth, R.E. Rose, and G. Skillman. 1987. *Oregon Territorial Sea Management Study.* Oregon Sea Grant College Program, Oregon State University, Corvallis, OR.

Gordon, D.C., P. Schwinghammer, T.W. Rowell, J. Prena, K. Gilkinson, W.P. Vass, and D.L. McKeown. 1998. Studies in eastern Canada on the impact of mobile fishing gear on benthic habitat and communities. In *Effects of Fishing Gear on the Sea Floor of New England,* Dorsey, E.M. and J. Pederson (eds.). Conservation Law Foundation, Boston, MA.

Greene, H.G., M.M. Yoklavich, R.M. Starr, V.M. O'Connell, W.W. Wakefield, D.E. Sullivan, J.E. McRea, Jr., and G.M. Cailliet. 1999. A classification scheme for deep seafloor habitats. *Oceanologica Acta* 22(6):663-678.

Greenstreet, S.P.R. and S.J. Hall. 1996. Fishing and groundfish assemblage structure in the northwestern North Sea: An analysis of long-term and spatial trends. *Journal of Animal Ecology* 65:577-598.

Guillen, J.E., A.A. Ramos, L. Martinez, and J. Sanchez Lizaso. 1994. Anti-trawling reefs and the protection of Posidonia Oceanica Meadows in the western Mediterranean Sea: demands and aims. *Bulletin of Marine Science* 55(2-3):645-650.

Gulland, J.A. 1983. *Fish Stock Assessment: A Manual of Basic Methods.* John Wiley and Sons, New York, 223pp.

Hall-Spencer, J.M. and P.G. Moore. 2000. Impact of scallop dredging on maerl grounds. In *Effects of Fishing on Non-Target Species and Habitats,* Kaiser, M.J. and S.J. de Groot (eds.). Blackwell, Oxford, UK.

Herrmann, M., J. Greenberg, and K. Criddle. 1998. An economic analysis of pot limits for the Adak brown king crab fishery: A distinction between open access and common property. *Alaska Fishery Research Bulletin* 5(1):25-38.

Hiatt, T. and J. Terry. 1999. *Stock Assessment and Fishery Evaluation Report for the Groundfish Fisheries of the Gulf of Alaska and Bering Sea/Aleutian Islands Area: Economic Status of the Groundfish Fisher-*

ies off Alaska, 1999. Seattle, WA: Socioeconomic Assessments Program, Resource Ecology and Fisheries Management Division, Alaska Fisheries Science Center, National Oceanic and Atmospheric Administration.

Hiatt, T., R. Felthoven, and J. Terry. 2001. *Stock Assessment and Fishery Evaluation Report for the Groundfish Fisheries of the Gulf of Alaska and Bering Sea/Aleutian Islands Area: Economic Status of the Groundfish Fisheries off Alaska, 2000*. Socioeconomic Assessments Program, Resource Ecology and Fisheries Management Division, Alaska Fisheries Science Center, NOAA.

Holling, C.S. 1973. Resilience and stability of ecological systems. *Annual Review of Ecology and Systematics* 4:1-23.

Holling, C.S., D.W. Schindler, B. Walker, and J. Roughgarden. 1995. Biodiversity and the functioning of ecosystems: An ecological synthesis. In *Biodiversity Loss: Ecological and Economic Issues*, C. Perrings, K.G. Mäler, C. Folke, C.S. Holling, and B.O. Jansson (eds.). Cambridge University Press.

Hopkinson, C.S., A.E. Giblin, J. Tucker, and R.H. Garritt. 1999. Benthic metabolism and nutrient cycling along an estuarine salinity gradient. *Estuaries* 22:825-843.

Hutchings, J.A. 2000. Collapse and recovery of marine fishes. *Nature* 406:882-885.

Ianelli, J.N. and V. Wespestad. 1998. *Trends in North Pacific Cod and Pollock Catch 1981-1998*. Resource Ecology and Fisheries Management, NOAA/NMFS Alaska Fisheries Science Center.

International Council for the Exploration of the Sea. 1996. *Report of the Working Group on Ecosystem Effects of Fishing Activity*. ICES C.M. 1996/Assess/Env:1 Ref.:Session G.

International Council for the Exploration of the Sea. 1998. *Report of the Working Group on Ecosystem Effects of Fishing Activities*. ICES C.M. 1998/ACME:1.

International Council for the Exploration of the Sea. 2000. *Report of the Working Group on Ecosystem Effects of Fishing Activities*. ICES C.M. 2000/ACME:2.

International Council for the Exploration of the Sea. 2001. *Report of the Working Group on Ecosystem Effects of Fishing Activities*. ICES C.M. 2001/ACME:09.

Jackson, J.B.C., M.X. Kirby, W.H. Berger, K.A. Bjorndal, L.W. Botsford, B.J. Bourque, R.H. Bradbury, R. Cooke, J. Erlandson, J.A. Estes, T.P. Hughes, S. Kidwell, C.B. Lange, H.S. Lenihan, J.M. Pandolfi, C.H. Peterson, R.S. Steneck, M.J. Tegner, and R.R. Warner. 2001. Historical overfishing and the recent collapse of coastal ecosystems. *Science* 293:629-637.

Jennings, S. and M.J. Kaiser. 1988. The effects of fishing on marine ecosystems. In *Advances in Marine Biology 34*, Blaxter, J.H.S., A.J. Southward, and P.A. Tyler (eds.). Academic Press.

Jennings, S., J.K. Pinnegar, N.V.C. Polunin, and K.J. Warr. 2001. Impacts of trawling disturbance on the trophic structure of benthic invertebrate communities. *Marine Ecology Progress Series* 213:127-142.

Kaiser, M.J. 1998. Significance of bottom-fishing disturbance. *Conservation Biology* 12(6):1230-1235.

Kaiser, M.J. and B.E. Spencer. 1996a. Behavioural responses of scavengers to beam trawl disturbance. In *Aquatic Predators and Their Prey*, Greenstreet, S.P.R. and M.L. Tasker (eds.). Blackwell Scientific Publications, Oxford.

Kaiser, M.J. and B.E. Spencer. 1996b. The effects of beam-trawl disturbance on infaunal communities in different habitats. *Journal of Animal Ecology* 65(3):348-358.

Kaiser, M.J, S.I. Rogers, and J.R. Ellis. 1999. Importance of benthic habitat complexity for demersal fish assemblages. *American Fisheries Society Symposium* 22:212-223.

Kaiser, M.J., K. Ramsay, C.A. Richardson, F.E. Spence, and A.R. Brand. 2000. Chronic fishing disturbance has changed shelf sea benthic community structure. *Journal of Animal Ecology* 69(3):494-503.

Kemp, W.M., W.R. Boynton, R.R. Twilley, J.C. Stevenson, and J.C. Mean. 1983. The decline of submerged vascular plants in upper Chesapeake Bay: Summary of results concerning possible causes. *Marine Technology Society of Japan* 17:78-89.

Kenchington, E.L.R., J. Prena, K.D. Gilkinson, D.C. Gordon, Jr., K. MacIsaac, C. Bourbonnais, P.J. Schwinghamer, T.W. Rowell, D.L. McKeown, and W.P. Vass. 2001. Effects of experimental otter trawling on the macrofauna of a sandy bottom ecosystem on the Grand Banks of Newfoundland. *Canadian Journal of Fisheries and Aquatic Sciences* 58:1043-1057.

Knowlton, N. 1992. Threshold and multiple states in coral reef community dynamics. *American Zoologist* 32:674-682.

Koenig, C.C., F.C. Coleman, C.B. Grimes, G.R. Fitzhugh, K.M. Scanlon, C.T. Gledhill, and M. Grace. 2000. Protection of fish spawning habitat for the conservation of warm-temperature reef-fish fisheries of shelf-edge reefs of Florida. *Bulletin of Marine Science* 66(3):593-616.

Kostylev, V.E., R.C. Courtney, G. Robert, and B.J. Todd. In press. Stock evaluation of giant scallop *(Placopecten magellanicus)* using high-resolution acoustics. *Fisheries Research*.

Lake, P.S. 1990. Disturbing hard and soft bottom communities: A comparison of marine and freshwater environments. *Australian Journal of Ecology* 15:477-489.

Langton, R.W., P.J. Auster, and D.C. Schneider. 1995. A spatial and temporal perspective on research and management of groundfish in the northwest Atlantic. *Reviews in Fisheries Science* 3:201-229.

Leet, W.S., C.M. Dewees, and C.W. Haugen (eds.). 1992. *California's Living Marine Resources and Their Utilization*. Sea Grant Extension Program, Department of Wildlife and Fisheries Biology, University of California.

Lenihan, H.S. and C.H. Peterson. 1998. How habitat degradation through fishery disturbance enhances impacts of hypoxia on oyster reefs. *Ecological Applications* 8(1):128-140.

Lenihan, H.S., C.H. Peterson, J.E. Byers, J.H. Grabowski, G.W. Thayer, and D.R. Colby. 2001. Cascading of habitat degradation: Oyster reefs invaded by refugee fishes escaping stress. *Ecological Applications* 11(3):764-782.

Lester, J. 1995. *Report of the Ecosystems Subpanel*. Wilson, J., S. Strawn, and D. Hitchcock (eds.). Houston Advanced Research Center, TX.

LeVieil, D.P. 1987. *Territorial Use Rights in Fishing (TURFs) and the Management of Small-Scale Fishing: The Case of Lake Titicaca (Peru)*. Dissertation, University of British Columbia.

Leys, S.P. and N.R.J. Lauzon. 1998. Hexactinellid sponge ecology: Growth rates and seasonality in deep water sponges. *Journal of Experimental Marine Biology and Ecology* 230:111-129.

Lindegarth, M., D. Valentinsson, M. Hansson, and M. Ulmestrand. 2000. Effects of trawling disturbances on temporal and spatial structure of benthic soft-sediment assemblages in Gullmarsfjorden, Sweden. *ICES Journal of Marine Science* 57:1369-1376.

Lindholm, J.B., P.J. Auster, and L. Kaufman. 1999. Habitat mediated survivorship of 0-year Atlantic cod (*Gadus morhua*). *Marine Ecology Progress Series* 180:247-255.

Lindholm, J.B., P.J. Auster, M. Ruth, and L. Kaufman. 2001. Juvenile fish responses to variations in seafloor habitats: Modeling the effects of fishing and implications for the design of marine protected areas. *Conservation Biology* 15:424-437.

Magorrian, B.H. 1996. *The Impact of Commercial Trawling on the Benthos of Strangford Lough*. The Queen's University of Belfast, Northern Ireland.

Main, J. and G.I. Sangster. 1981. A study of the sand clouds produced by trawl boards and their possible effect on fish capture. *Scottish Fisheries Research Report No. 20*. Department of Agriculture and Fisheries for Scotland, Marine Laboratory, Aberdeen.

Manson, G. and B.J. Todd. 2000. Revolution in the Nova Scotia scallop fishery: Seabed maps turn hunting into harvesting. *Fishing News International* 39(2):20-22.

Mason, J. 1983. *Scallop and Queen Fishing in the British Isles.* Fishing News Books, London.

Mayer, L.M., D.F. Schick, R.H. Findaly, and D.L. Rice. 1991. Effects of commercial dragging on sedimentary organic matter. *Marine Environmental Research* 31:249-261.

McCay, B.J. and M. Cieri. 2000. *2001 Summer Flounder, Scup, and Black Sea Bass Specifications. Environmental Assessment.* Mid-Atlantic Fishery Management Council.

McCay, B., and C. Creed. 1990. Social structure and debates on fisheries management in the Atlantic surf clam fishery. *Ocean and Shoreline Management* 13:199-229.

Meyer, D.L., M.S. Fonseca, P.L. Murphy, R.H. McMichael, Jr., M.M. Byerly, M.W. LaCroix, P.E. Whitfield, and G.W. Thayer. 1999. Effects of live-bait shrimp trawling on seagrass beds and fish bycatch in Tampa Bay, Florida. *Fishery Bulletin* 97(1):193-199.

Moore, D.R. and H.R. Bullis, Jr. 1960. A deep water coral reef in the Gulf of Mexico. *Bulletin of Marine Science of the Gulf and Caribbean* 10(1):125-128.

Moore, K.A., H.A. Neckles, and R.J. Orth. 1996. *Zostera marina* (eelgrass) growth and survival along a gradient of nutrients and turbidity in the lower Chesapeake Bay. *Marine Ecology Progress Series* 142:247-259.

Moore, K.A. and R.J. Orth. 1997. *Evidence of Widespread Destruction of Submersed Aquatic Vegetation (SAV) from Clam Dredging in Chincoteague Bay, Virginia,* [Online]. Available: http://www.vims.edu/bio/sav/clamdredge/ [2002, March 29].

Murawski, S.A., R. Brown, H.L. Lai, P.J. Rago, and L. Hendrickson. 2000. Large-scale closed areas as a fishery-management tool in temperate marine systems: The Georges Bank experience. *Bulletin of Marine Science* 66(3):775-798.

National Geophysical Data Center. 2002. *Bathymetry, Topography, and Relief.* [Online]. Available: http://www.ngdc.noaa.gov/mgg/bathymetry/relief.html [2002, February 22].

National Marine Fisheries Service. 1997. Magnuson-Stevens Act provisions: Essential fish habitat. U.S. Department of Commerce, National Oceanic and Atmospheric Administration, Interim Final Rule. *Federal Register* 62(244):66,531-66,559.

National Marine Fisheries Service. 2001a. *Alaska Groundfish Fisheries: Draft Programmatic Supplemental Environment Impact Statement.* National Marine Fisheries Service, Alaska Region, National Oceanic and Atmospheric Administration, U.S. Department of Commerce.

National Marine Fisheries Service. 2001b. *Fishery Economic Trends,* [Online]. Available: http://www.nefsc.nmfs.gov/sos/econ [2001, October 8].

National Marine Fisheries Service. 2001c. *Appendix C: Regulatory Impact Review of Steller Sea Lion Protection Measures.* Draft Supplemental Environmental Impact Statement. National Marine Fisheries Service, Alaska Region, Juneau.

National Marine Fisheries Service. 2002. Magnuson-Stevens Act provisions: Essential fish habitat. *Federal Register* 67(12):2343-2383.

National Oceanic and Atmospheric Administration. 1999. *Our Living Oceans.* U.S. Department of Commerce.

National Research Council. 1983. *Risk Assessment in the Federal Government: Managing the Process.* National Academy Press, Washington, D.C.

National Research Council. 1993. *Issues in Risk Assessment.* National Academy Press, Washington, D.C.

National Research Council. 1994. *Improving the Management of U.S. Marine Fisheries.* National Academy Press, Washington, D.C.

National Research Council. 1996. *Understanding Risk: Informing Decisions in a Democratic Society.* National Academy Press, Washington, D.C.

National Research Council. 1999. *Sharing the Fish: Toward a National Policy on Individual Fishing Quotas.* National Academy Press, Washington, D.C.

National Research Council. 2001. *Marine Protected Areas: Tools for Sustaining Ocean Ecosystems.* National Academy Press, Washington, D.C.

Natural Resource Consultants. 1997. *Bottom Contact Fisheries of Onslow Bay, North Carolina.* Seattle, WA.

Natural Resource Consultants. 1999. *Status of Washington Based Commercial Fisheries and Fleets Future Utilization of Fishermen's Terminal.* Prepared for Port of Seattle, Fishermen's Terminal Market Research Project. Seattle, WA.

New England Fishery Management Council. 2000. *Atlantic Sea Scallop Fishery Management Plan Stock Assessment and Fishery Evaluation Report,* [Online]. Available: http://www.nefmc.org/ [2001, August 30].

New England Fishery Management Council. 2002. *Council Report.* NEFMC, Newburyport, MA.

North Pacific Fishery Management Council. 1999. *Environmental Assessment/Regulatory Impact Review/Initial Regulatory Flexibility Analysis for Amendment 57 to the FMP for the Groundfish Fishery of the Bering Sea and Aleutian Islands Area to Prohibit the Use of Nonpelagic Trawl Gear in Directed Pollock Fisheries.* North Pacific Fishery Management Council, Draft for Secretarial Review, Anchorage, AK.

O'Connell, V.M., W. Wakefield, and H. Greene. 1998. The use of a no-take marine reserve in the eastern Gulf of Alaska to protect essential fish habitat. In *Marine Harvest Refugia for West Coast Rockfish: A Workshop,* Yoklavich, M. (ed.). NOAA Technical Memorandum NMFS-SWFSC-255.

Odum, W.E. 1982. Environmental degradation and the tyranny of small decisions. *BioScience* 32:728-729.

Pacific Fishery Management Council. 2000. *Overcapitalization in the West Coast Groundfish Fishery: Background, Issues, and Solutions.* Economic Subcommittee, Scientific and Statistical Committee, Portland, OR.

Pacific States Marine Fisheries Commission. 1998. *West Coast Catcher Boat Survey Summary 1997-1998.* Economic Data Program, Pacific States Marine Fisheries Commission, Seattle, WA.

Pacific States Marine Fisheries Commission. 1999. *52nd Annual Report of the Pacific States Marine Fisheries Commission for the Year 1999.* Pacific States Marine Fisheries Commission, Seattle, WA.

Patten, B.C. and R. Constanza. 1997. Logical interrelations between four sustainability parameters: Stability, continuation, longevity, and health. *Ecosystem Health* 3:136-142.

Peterson, C.H., H.C. Summerson, and S.R. Fegley. 1987. Ecological consequences of mechanical harvesting of clams. *Fishery Bulletin* 85:281-298.

Phillips, G.L., D. Eminson, and B. Moss. 1978. A mechanism to account for macrophyte decline in progressively eutrophicated freshwaters. *Aquatic Botany* 4:103-126.

Pickett, S.T.A. and P.S. White. 1995. *The Ecology of Natural Disturbance and Patch Dynamics.* Academic Press, New York, NY.

Pilskaln, C.H., J.H. Churchill, and L.M. Mayer. 1998. Resuspension of sediment by bottom trawling in the Gulf of Maine and potential geochemical consequences. *Conservation Biology* 12(6):1223-1224.

Poiner, I., J. Glaister, R. Pitcher, C. Burridge, T. Wassenberg, N. Gribble, B. Hill, S. Blaber, D. Milton, D. Brewer, and N. Ellis. 1998. *Final Report on Effects of Trawling in the Far Northern Section of the Great Barrier Reef: 1991-1996.* CSIRO Division of Marine Research, Cleveland, OH.

Poppe, L.J. and C.F. Polloni. 2000. *USGS East-Coast Sediment Analysis: Procedures, Database, and Geo-referenced Displays.* U.S. Geological Survey Open-File Report 00-358.

Presidential/Congressional Commission on Risk Assessment and Risk Management. 1997. *Risk Assessment and Risk Management in Regulatory Decision-Making,* [Online]. Available: http://www.epa.gov/ncea/pres_com/riskcom2/v2epaa.htm [2001, August 30].

Radtke, H.D. and S.W. Davis. 2000. *Description of the U.S. West Coast Commercial Fishing Fleet and Seafood Processors.* Prepared for the

Pacific States Marine Fisheries Commission. The Research Group, Corvallis, OR.

Reynolds, J.R., R.C. Highsmith, B. Konar, C.G. Wheat, and D. Doudna. 2001. *Fisheries and Fisheries Habitat Investigations Using Undersea Technology*, [Online]. Available: http://www.wcnurc.uaf.edu:8000/ [2002, February 1].

Riemann, B. and E. Hoffmann. 1991. Ecological consequences of dredging and bottom trawling in the Limfjord, Denmark. *Marine Ecology Progress Series* 69:171-178.

Rjinsdorp, A.D., A.M. Buijs, F. Storbeck, and E. Visser. 1998. Microscale distribution of beam trawl effort in the southern North Sea between 1993 and 1996 in relation to the trawling frequency of the sea bed and the impact on benthic organisms. *ICES Journal of Marine Science* 55:403-419.

Roff, J.C. and M.E. Taylor. 2000. National frameworks for marine conservation: A hierarchical geophysical approach. *Aquatic Conservation: Marine and Freshwater Ecosystems* 10:209-223.

Sainsbury, K.J., R.A. Campbell, R. Lindholm, and A.W. Whitelaw. 1997. Experimental management of an Australian multispecies fishery: Examining the possibility of trawl induced habitat modification. In *Global Trends: Fisheries Management*, Piktch, E.K., D.D. Huppert, and M.P. Sissenwine (eds.). American Fisheries Society, Symposium 20, Bethesda, MD.

Sale, P.F. (ed.). 1991. *The Ecology of Fishes on Coral Reefs*. Academic Press, New York.

Schlee, J., D.W. Folger, and C.J. O'Hara. 1973. Bottom sediments on the continental shelf off the northeastern United States: Cape Cod to Cape Ann, Massachusetts. Miscellaneous Geologic Investigations Map I-749, U.S. Geological Survey, Washington, D.C.

Schwinghamer, P, D.C. Gordon, T.W. Rowell, J. Prena, D.L. McKeown, G. Sonnichsen, and J.Y. Guigne. 1998. Effects of experimental otter trawling on surficial sediment properties of a sandy-bottom ecosystem on the Grand Banks of Newfoundland. *Conservation Biology* 12:215-1222.

Shapiro, S. (ed.). 1971. *Our Changing Fisheries*. National Oceanic and Atmospheric Administration. U.S. Department of Commerce.

Sheridan, P.F. 2001. *Pete Sheridan Presentation Notes*. Presentation to the Committee on Ecosystem Effects of Fishing: Phase 1—Effects of Bottom Trawling on Seafloor Habitat, Galveston, TX, March 2001.

Short, F.T, D.M. Burdick, and J.E. Kaldy, III. 1995. Mesocosm experiments quantify the effects of eutrophication on eelgrass, *Zostera marina*. *Limnology and Oceanography* 40(4):740-749.

Smith, E.M., M.A. Alexander, M.M. Blake, L. Gunn, P.T. Howell, M.W. Johnson, R.E. MacLeod, R.F. Sampson, Jr., D.G. Simpson, W.H. Webb, L.L. Stewart, P.J. Auster, N.K. Bender, K. Buchholz, J. Crawford, and T.J. Visel. 1985. *A Study of Lobster Fisheries in the Connecticut Waters of Long Island Sound with Reference to the Effects of Trawling on Lobsters*. Connecticut Department of Environmental Protection Marine Fisheries Program, Hartford.

Smith, S.J., J.J. Hunt, and D. Rivard (eds.). 1993. *Risk Evaluation and Biological Reference Points for Fisheries Management*. Canadian Special Publication of Fisheries and Aquatic Sciences.

Smolowitz, R. 1998. Bottom tending gear used in New England. In *Effects of Fishing Gear on the Sea Floor of New England*, Dorsey, E.M. and J. Pederson (eds.). Conservation Law Foundation, Boston, MA.

South Atlantic Fishery Management Council. 1996. *Final Amendment 2 (Bycatch Reduction) to the Fishery Management Plan for the Shrimp Fishery of the South Atlantic Region, April 1996*. South Atlantic Fishery Management Council, Charleston, South Carolina.

South Atlantic Fishery Management Council. 2000. *Revised Final Fishery Management Plan for the Calico Scallop Fishery in the South Atlantic Region, August 2000*. South Atlantic Fishery Management Council, Charleston, SC.

South Atlantic Fishery Management Council. 2001. *Public Hearing Draft Amendment 5 to the Fishery Management Plan for the Shrimp Fishery of the South Atlantic Region (Rock Shrimp), April 2001*. South Atlantic Fishery Management Council, Charleston, SC.

Starr, R.M., K.A. Johnson, E.A Laman, and G.M. Cailliet. 1998. *Fishery Resources of the Monterey Bay National Marine Sanctuary*. California Sea Grant College System, University of California, La Jolla, CA.

Stein, D.L., B.N. Tissot, M.A. Hixon, and W. Barss. 1992. Fish-habitat associations on a deep reef at the edge of the Oregon continental shelf. *Fishery Bulletin* 90:540-551.

Stephan, C.D., R.L. Peuser, and M.S. Fonseca. 2000. Evaluating fishing gear impacts to submerged aquatic vegetation and determining mitigation strategies. *Atlantic States Marine Fisheries Commission Habitat Management Series #5*. 38pp.

Thayer, G.W., W.J. Kenworth, and M.S. Fonesca. 1984. *The Ecology of Eelgrass Meadows of the Atlantic Coast: A Community Profile*. U.S. Department of the Interior, Washington, D.C.

Thomas, J.S., G.D. Johnson, C.M. Formichella, and C.A. Riordan. 1995. *Gulf Fishermen on the Eve of Bycatch Regulations*. Marfin #NA37FF0049.

Thomson, C. 2001. The human ecosystem. In *California's Living Marine Resources: A Status Report*, Leet, W.S., R. Klingbell, C.M. Dewees, and E. Larson (eds.). California Department of Fish and Game, Sacramento, CA.

Thrush, S.F., J.E. Hewitt, V.J. Cummings, P.K. Dayton, M. Cryer, S.J. Turner, G.A. Funnell, R.G. Budd, C.J. Milburn, and M.R. Wilkinson. 1998. Disturbance of the marine benthic habitat by commercial fishing: impacts at the scale of the fishery. *Ecological Applications* 8(3):866-879.

Tilmant, J.T. 1979. Observations on the impact of shrimp roller frame trawls operated over hard-bottom assemblage of sponges and corals. *Fisheries Research* 5:39-54.

Tuck, I.D., S.J. Hall, M.R. Robertson, E. Armstrong, and D.J. Basford. 1998. Effects of physical trawling disturbance in a previously unfished sheltered Scottish sea loch. *Marine Ecology Progress Series* 162:227-242.

Tuck, I.D., N. Bailey, M. Harding, G. Sangster, T. Howell, N. Graham, and M. Breen. 2000. The impact of water jet dredging for razor clams, *Ensis* sp., in a shallow sandy subtidal environment. *Journal of Sea Research* 43:65-81.

Tupper, M. and R.G. Boutilier. 1995. Effects of habitat on settlement, growth, and postsettlement survival of Atlantic cod (*Gadus morhua*). *Canadian Journal of Fisheries and Aquatic Sciences* 52:1834-1841.

Twilley, R.R., W.M. Kemp, K.W. Staver, J.C. Stevenson, and W.R. Boynton. 1985. Nutrient enrichment of estuarine submerged vascular plant communities. 1. Algal growth and effects on production of plants and associated communities. *Marine Ecology Progress Series* 23:179-191.

Ulanowicz, R.E. and J.H. Tuttle. 1992. The trophic consequences of oyster stock rehabilitation in Chesapeake Bay. *Estuaries* 15(3):298-306.

U.S. Geological Survey. 2001. *Sea-Floor Mapping: Data Acquisition*, [Online]. Available: http://woodshole.er.usgs.gov/operations/sfmapping/dataacq.htm [2002, February 22].

Valentine, P.C., T.J. Middleton, and S.J. Fuller. 2001. Seafloor maps showing topography, sun-illuminated topography, and backscatter intensity of the Stellwagen Bank National Marine Sanctuary region off Boston, Massachusetts. *U.S. Geological Survey Open-File Report* 00-410.

Valentine, P.C., T.S. Unger, and J.L. Baker. In press. *Sedimentary Environments of Quadrangle 6 in the Stellwagen Bank National Marine Sanctuary off Boston, Massachusetts*. U.S. Geological Survey Geologic Investigations Series.

Van Dolah, R.F., P.H. Wendt, and N. Nicholson. 1987. Effects of a research trawl on a hard bottom assemblage of sponges and corals. *Fisheries Research* 5:39-54.

Wakefield, W.W., V.M. O'Connell, H.G. Greene, D.W. Carlisle, and J.E. McRea, Jr. 1998. The role of sidescan sonar in seafloor classification with a direct application to commercial fisheries management. In

Deepwater Fish and Fisheries. International Council for the Exploration of the Sea, Copenhagen, Denmark.

Walters, C.J. 1986. *Adaptive Management of Renewable Resources.* Macmillan, New York, NY.

Walters, C.J. and F. Juanes. 1993. Recruitment limitation as a consequence of natural selection for use of restricted feeding habitats and predation risk taking by juvenile fishes. *Canadian Journal of Fisheries and Aquatic Sciences* 50:2058-2070.

Watling, L. and E.A. Norse. 1998. Disturbance of the seabed by mobile fishing gear: A comparison to forest clearcutting. *Conservation Biology* 12(6):1180-1197.

Yoklavich, M., H.G. Greene, G. Cailliet, D. Sullivan, R. Lea, and M. Love. 2000. Habitat associations of deep-water rockfishes in a submarine canyon: An example of a natural refuge. *Fishery Bulletin* 98:625-641.

Appendixes

Appendix A

Committee and Staff Biographies

COMMITTEE

John Steele *(Chair)* has worked as a Senior Scientist at the Woods Hole Oceanographic Institution since 1977 (Director, 1977–1989). Dr. Steele earned a D.Sc. in biology in 1964 from University College, London. His research focuses on marine ecosystem dynamics. He is a former member of the Ocean Studies Board. He received the National Academy of Sciences Alexander Agassiz Medal in 1973 and is a Fellow of the Royal Society.

Dayton Lee Alverson worked for the Washington Department of Fisheries, Bureau of Commercial Fisheries, and National Marine Fisheries Service during his career as a fisheries research scientist. After his retirement from government, he formed Natural Resources Consultants and served as its President and Chairman of the Board. He has recently stepped down from the Chairman's position and currently serves the company as Senior Scientist. Dr. Alverson earned a Ph.D. in fisheries and oceanographic science in 1967. His research has focused on resource surveys and assessment, fish behavior, status of marine stocks, bycatch, natural resource policy, and fishing gear technology.

Peter Auster has worked as Science Director for the National Undersea Research Center at the University of Connecticut since 1992. Dr. Auster earned an M.S. in biological oceanography in 1985 from the University of Connecticut and a Ph.D. in zoology from the National University of Ireland in Galway in 2000. His research focuses on the effects of fishing gear on the environment, the role of habitat on the distribution and abundance of mobile fauna, the linkages between habitat-level processes and population-community dynamics, and the scientific basis for marine protected areas. He received a Pew fellowship in Marine Conservation.

Jeremy Collie has been a Professor of Oceanography at the University of Rhode Island since 2001. Dr. Collie earned a Ph.D. in biological oceanography in 1985 from the Massachusetts Institute of Technology and Woods Hole Oceanographic Institution joint program. His research focuses on the effects of disturbance on benthic communities, quantitative ecology with emphasis on population dynamics and production of marine mammals, fish population dynamics and management, and predator-prey interactions.

Joseph T. DeAlteris has been a Professor at the University of Rhode Island since 1995. Dr. DeAlteris earned a Ph.D. in 1986 from the Virginia Institute of Marine Science, with a specialization in physical processes. His recent research has focused on aquatic resource harvesting technologies and their effect on the ecosystem, in particular the reduction of bycatch through development of size- and species-specific fishing gear and the quantitative evaluation of effects of fishing gear on fish stocks, habitat, and manmade structures placed on and under the seabed.

Linda Deegan has worked as an Associate Scientist at the Marine Biological Laboratory in Woods Hole since 1989. Dr. Deegan earned a Ph.D. in marine sciences in 1985 from the Louisiana State University. Her research focuses on the relationship between ecosystem dynamics and animal populations, effects of habitat degradation on fish community structure, and the importance of fish in exporting nutrients and carbon from estuaries.

Elva Escobar-Briones has worked at the Instituto de Ciencias del Mar y Limnologia at the Universidad Nacional Autonoma de Mexico since 1989. Dr. Escobar-Briones earned a Ph.D. in biological oceanography in 1987 from the Universidad Nacional Autonoma de Mexico. Her research focuses on benthic community structure and function. She served as a committee member of the Ocean Studies Board's Academia Mexicana de Ciencias/National Research Council joint working group on ocean sciences. She has served as head of the Department of Coastal and Ocean Sciences since 1999.

Stephen J. Hall was Professor of Marine Biology at Flinders University of South Australia before his recent appointment as Director of the Australian Institute of Marine Science. He has published extensively on the structure and functioning of marine ecological systems, focusing especially on the effects of natural and human disturbance. He has recently published a book on the global effects of fishing on marine communities and ecosystems. Dr. Hall has served on numerous national and international committees and is a past chairman of the International Council for the Exploration of the Seas (ICES) Working Group on the Ecosystem Effects of Fishing Activities—a group that provides advice to ICES and the European Commission on fishing effects and other aspects of coastal zone management. He is also a recent recipient of a Pew Fellowship in Marine Conservation.

Gordon H. Kruse has worked as the President's Professor of Fisheries with the University of Alaska Fairbanks, Juneau Center, School of Fisheries and Ocean Sciences since November 2001. Previously, Dr. Kruse worked for the Alaska Department of Fish and Game. In his capacity as Chief Marine Fisheries Scientist he led the state's marine fishery research program for shellfish, groundfish, and herring. Dr. Kruse's marine fisheries research interests include population estimation models, stock production parameters, fishery management strategies, population and ecosystem dynamics, and fishery oceanography. He earned a Ph.D. in fisheries science in 1983 from Oregon State University. Dr. Kruse is U.S. delegate to the Fishery

Science Committee of the North Pacific Marine Science Organization, and he serves as a member of the Committee on the Alaska Groundfish Fishery and Steller Sea Lions.

Caroline Pomeroy has been an Assistant Research Scientist at the University of California, Santa Cruz, since 1995. Dr. Pomeroy earned a M.A. in marine affairs and policy from the University of Miami Rosenstiel School in 1989 and a Ph.D. in wildlife and fisheries science (with an emphasis in sociology) in 1993 from Texas A&M University. Her research focuses on the social, cultural, and economic aspects of fisheries and their management; the human dimensions of marine protected areas; and on the role of resource users and local knowledge in resource management. She also serves on the Research Activity Panel for the Monterey Bay National Marine Sanctuary.

Kathryn M. Scanlon is a Geologist with the U.S. Geological Survey's Coastal and Marine Geology Program. Ms. Scanlon earned a B.S. in geology in 1976 from Cornell University and an M.S. in geology in 1979 from the State University of New York at Albany. Her research focuses on mapping surficial seafloor geology, interpreting the geologic history of marine environments, and understanding relationships between biological communities and geologic processes in benthic marine habitats.

Priscilla Weeks has worked as a Research Associate at the Environmental Institute of Houston, University of Houston, Clear Lake, since 1993. Dr. Weeks earned a Ph.D. in anthropology in 1988 from the Rice University. Her research focuses on cultural anthropology, social aspects of natural resource management and environmental regulations, social aspects of natural resource management and rural development, and cross-cultural scientific collaboration. She is also a member of the socioeconomic panel that provides advice to the Gulf of Mexico Fishery Management Council.

STAFF

Susan Roberts serves as a Senior Program Officer with the Ocean Studies Board. Dr. Roberts received a Ph.D. in marine biology from the Scripps Institution of Oceanography. Her research interests include marine microbiology, fish physiology, and biomedicine. She staffs studies on the management of living marine resources.

Jodi Bachim serves as a Senior Project Assistant for the Ocean Studies Board. She received a B.S. in zoology from the University of Wisconsin-Madison in 1998. Since starting with the Ocean Studies Board in May 1999, Ms. Bachim has worked on several studies regarding fisheries, geology, nutrient over-enrichment, and marine mammals. Currently, she is working towards an M.S. in environmental biology.

Abby Schneider served as a National Research Council Fellow during the spring term of 2001. She received a B.S. in environmental engineering from the Massachusetts Institute of Technology and is currently enrolled in a Ph.D. program in environmental chemistry at the University of Maryland.

Appendix B

Regional Distribution of Fishing Effort

Commercial fisheries using trawls for groundfish and shrimp and dredges for scallops take place in the 6 fishery management regions of the contiguous United States. The information collected about those fisheries by the National Marine Fisheries Service (NMFS) and state agencies varies in quality and content from one region to another. Relatively good qualitative information is available for bottom trawl and scallop dredge fisheries, and detailed information on the location of bottom trawling, the duration of tows and the associated catches is available for most of the intensively trawled areas. However, the methods of data collection, the details maintained, and the geographic resolution differ. Four of the six regions have sufficient information about the distribution of trawling effort during the 1990s to allow depiction of fishing effort by geographic statistical area. The material in this appendix was collected and prepared by Natural Resources Consultants. It appears in this report with their permission.

ATLANTIC SEABOARD (FIGURES B.1–B.5)

Information about the trawl effort off New England and the mid-Atlantic states was provided to the committee by the Northeast Fisheries Science Center. Between 1964 and 1993 catch and fishing effort data

were collected through a voluntary program using NMFS port agents in various fishing communities from Maine to Virginia. The system relied on interviews with captains of a portion of the fleet. Additional data were collected from weigh-out (dealer) transactions, auction sales, and from trucking companies. The objective of the program was to interview a large portion of the offshore-vessel operators, because their vessels were likely to fish in different areas from one trip to the next (or even on the same trip). Inshore, fixed-gear fleets that were more likely to fish the same grounds were thus interviewed less frequently.

For each vessel's trip interview, the port agent assigned a 10-minute square (10′ latitude by 10′ longitude) that best characterized the location of landings from the trip. When data were available, the trip landings were split into several 10-minute squares, particularly if the catches of various species changed during the trip. All non-interviewed trips were assigned to the most appropriate quarter-degree square, based on the vessels known fishing patterns, port landings, or vessel location, as supplied by captains. Thus, the effort database is a combination of interviewed trips wherein data are located to 10-minute squares and noninterviewed trips wherein the data were assigned to quarter-degree squares. Effort data are defined in 24-hour days, not including steaming time.

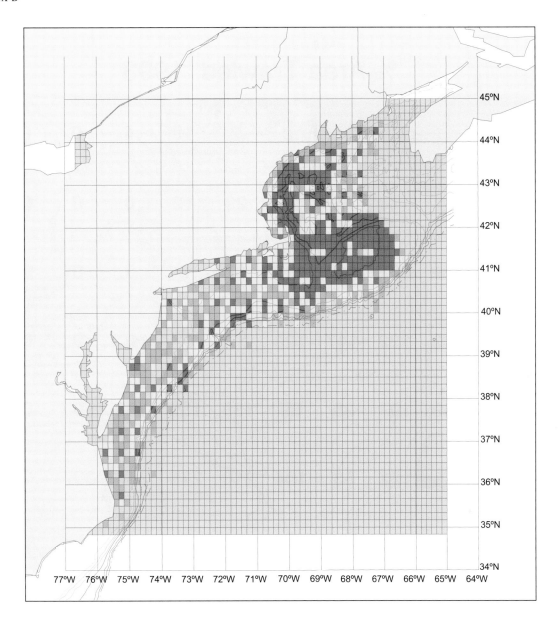

Number Days Fished off the Atlantic Seaboard
1991-1993

85 to 3982 (299)
20 to 85 (299)
1 to 19 (276)

FIGURE B.1 Distribution of bottom trawl effort in New England and mid-Atlantic regions, 1991–1993. After Canada extended its fisheries jurisdiction to 200 miles, the U.S. trawl fisheries of the region were limited to waters south of 44°30′N latitude. Fishing off the mid-Atlantic states, except at the northern edge, is directed at shrimp and occurs relatively close to shore at depths <100 m. Effort is measured as the number of days fished.

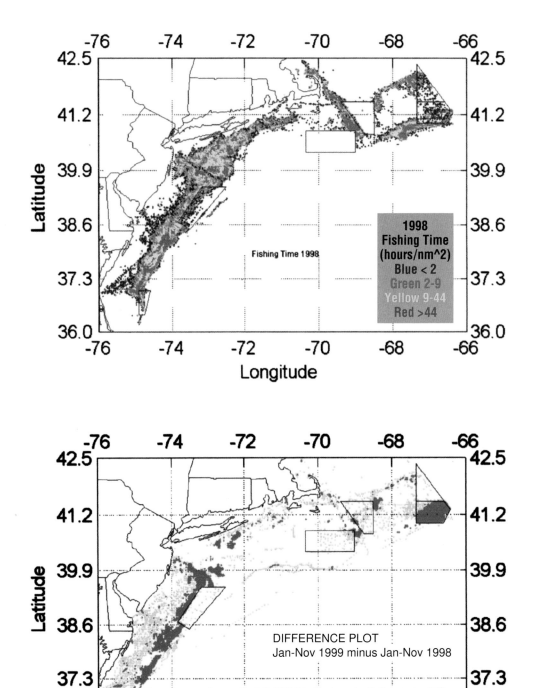

FIGURES B.4 and B.5 Scallop fishing effort off New England and the mid-Atlantic Bight, 1998 and 1999. These figures illustrate the distribution of scallop fishing effort for the years 1998 and 1999 in the New England and mid-Atlantic Bight regions. Although extensive areas of the region were dredged for scallops, the effort was considerably less than the bottom trawl fisheries. Recent data (1999–2000) for the region were not available in a similar format.

SOUTH ATLANTIC COAST (FIGURES B.6–B.7)

Fishery data along the southeastern coast of the United States has been collected by different agencies for the past 100 years. NMFS has been involved in fisheries data collection in the region for a considerable period, either as the sole entity, or in cooperation with state agencies. Unfortunately, because of the lack of effective regulations, states have been handicapped and trawl effort data in many instances are not very quantitative. Current estimates of the bottom trawling effort are patchy. Only two states, Florida and North Carolina, have implemented the use of trip tickets, thus improving the collection of effort data. Those two states, along with most of the Atlantic coast states, are developing a coast-wide, unified data collection system. In 1998–1999, there were 15,067 trawl trips by 901 licensed vessels reported off North Carolina.

In recent years more than 900 vessels have been involved in the North Carolina trawl fishery. About 82 percent of shrimp trawl trips during 1994–1997 occurred in estuaries. Occasional catches of shrimp are taken in the ocean areas and offshore vessels land close to a quarter of the state's shrimp landings. Most of the shrimp and crab trawling along the North Carolina coast occurs close to the beach, generally at depths <18.3 m.

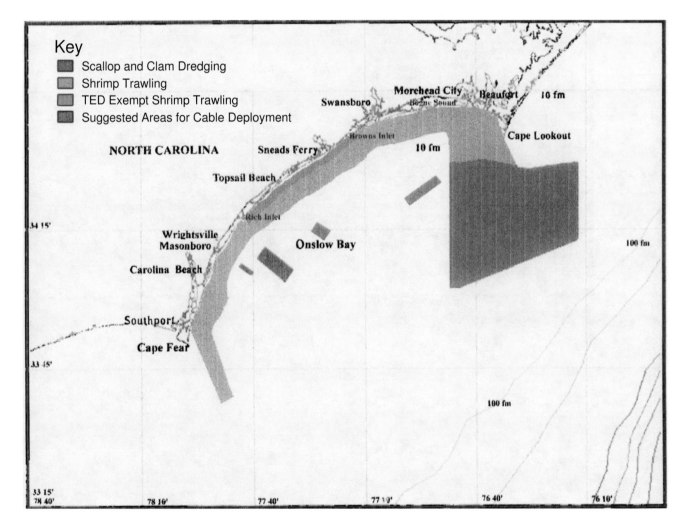

FIGURE B.6 Shrimp trawling and scallop and clam dredging in Onslow Bay. The distribution of trawling, scallop and clam dredging in Onslow Bay, North Carolina, is characteristic of the inshore trawl fisheries along the southeastern Atlantic region. An area between Rich's Inlet and Brown's Inlet extending 1 nm offshore is particularly heavily fished because of a special management opportunity. In 1995, 229 shrimp trawlers completed 3190 trips and covered 9570 nm^2 of the seabed, about four times the available trawl area (Natural Resources Consultants, 1997).

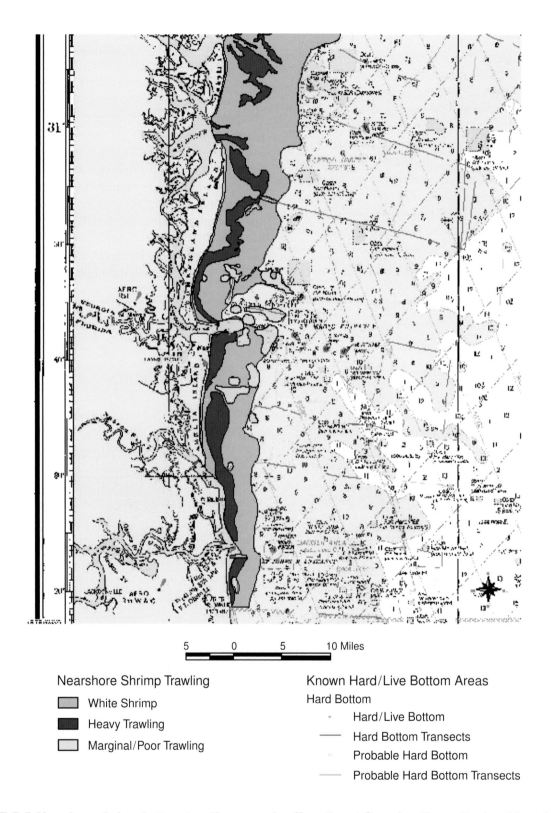

5   0   5   10 Miles

Nearshore Shrimp Trawling
- White Shrimp
- Heavy Trawling
- Marginal/Poor Trawling

Known Hard/Live Bottom Areas
Hard Bottom
- Hard/Live Bottom
- Hard Bottom Transects
- Probable Hard Bottom
- Probable Hard Bottom Transects

FIGURE B.7 Nearshore shrimp bottom trawling grounds off southeast Georgia. Karma Dunlop (Natural Resources Consultants) interviewed fishermen in Georgia and developed charts showing the relative distribution of shrimp trawling off the southeastern portion of the state. The figure illustrates the proximity of the trawl effort to the shoreline; most of the fishing occurs within four miles of the shoreline. Trawling off South Carolina also is carried out close to shore. During 1999, about 530 vessels were licensed to fish off Georgia.

GULF OF MEXICO (FIGURES B.8–B.10)

Although there has been small-scale trawl fishing for finfish for food and industrial purposes in the Gulf, essentially all bottom trawling now is directed at harvesting various species of shrimps. Scallop dredging and trawling also occurs in the region.

NMFS began collecting standardized fishery statistics for the Gulf in 1960 (catch, effort, gear, locale) using a shore-based sampling program, that more recently has been coupled with a low-level at-sea observer program. James M. Nance, NMFS Galveston Laboratory, provided the committee with detailed effort data for the Gulf of Mexico. Additional information was provided by Ecological Research Associates, Byran, Texas, a company involved in a voluntary sampling program that uses remote-sensing satellite positioning to track trawling in the Gulf throughout the year. NMFS notes four areas of shrimp bottom trawling concentration, as follows:

- Subarea 1–3, middle depths off Florida;
- Subarea 11, middle depths off Mississippi and Alabama;
- Subareas 13–17, shallow depths off Louisiana (west delta); and
- Subareas 18–19, shallow and mid-depths off west Texas.

Although the distribution of fishing effort can and does change daily, weekly, and monthly, annual averages over the 1990s did not vary much from one locality to another. Overall the bottom trawl effort declined 41 percent between the 1991–1993 and 1998–1999 periods. However, the Gulf of Mexico is still one of the most intensely bottom trawled areas off the United States. It has been noted (Sheridan, 2001) that the effort in some statistical blocks translates into a block area swept 37–75 times per year.

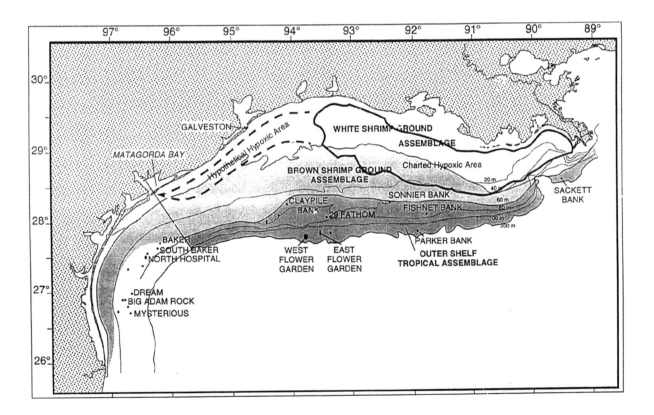

FIGURE B.8 Major shrimp grounds and faunal assemblages on continental shelf adjacent to Texas and Louisiana. This figure illustrates the major white and brown shrimp grounds, located primarily in the western Gulf, off the coasts of Louisiana and Texas. White shrimp typically are found in shallower nearshore waters.

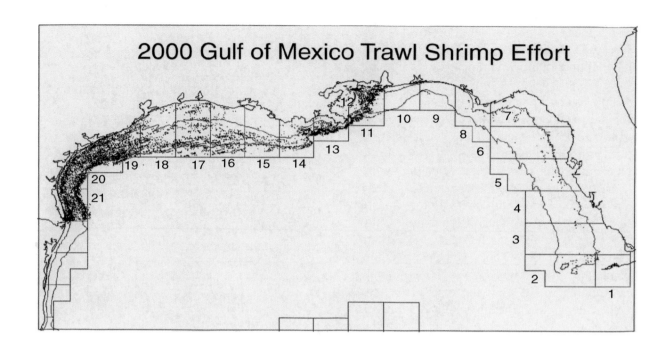

FIGURE B.9 Shrimp trawling effort distribution in the Gulf of Mexico. The distribution of shrimp trawling throughout the Gulf of Mexico in 2000 is based on passive "black box" positioning information obtained by Ecological Research Associates. The data were collected from vessel owners who volunteered to place positioning devices on their vessels. The graphic depiction represents only a small fraction of the total fishing effort. Nevertheless, the data provide a general picture of the major trawling areas.

Average Shrimp Trawl Fishing Days (24-hour) by Depth Contour (5 fathom intervals) in the Gulf of Mexico, 1991-1993

STATAREA	DEPTH00	DEPTH01	DEPTH02	DEPTH03	DEPTH04	DEPTH05	DEPTH06	DEPTH07	DEPTH08	DEPTH09	DEPTH10	STAT AREA
1	-	3.11	3.89	55.84	4.43	-	-	6.41	-	-	-	73.68
2	-	-	39.58	6,940.07	1,367.82	28.43	173.81	2.29	2.37	-	0.34	8,554.70
3	-	66.69	606.23	510.41	1,426.14	10.94	-	-	-	-	-	2,620.42
4	20.65	27.66	1,015.73	70.90	30.29	-	-	-	-	-	310.33	1,475.56
5	20.97	0.10	18.91	122.98	99.21	-	2.75	-	-	-	35.47	300.39
6	53.96	346.38	73.10	393.11	39.06	2.01	-	-	-	-	-	907.61
7	789.25	502.68	227.04	2,734.82	13.43	-	2.00	-	-	-	0.29	4,269.51
8	390.52	44.34	245.89	981.68	64.72	68.71	41.76	37.55	29.48	22.05	-	1,926.69
9	253.12	0.64	8.56	18.67	20.01	5.69	2.33	-	-	-	0.36	309.37
10	2,223.69	90.84	573.87	317.59	196.79	103.09	122.25	39.95	12.09	1.00	17.92	3,699.08
11	5,733.70	611.39	4,829.93	5,420.92	764.24	592.20	157.04	69.16	21.43	-	121.77	18,321.77
12	12,823.93	308.77	256.68	143.72	139.78	71.78	57.15	38.53	4.83	-	0.55	13,845.73
13	17,272.19	17,341.68	3,863.14	2,267.76	1,132.25	995.39	382.54	230.55	79.02	28.33	0.72	43,593.57
14	17,286.24	6,117.80	4,783.13	4,371.45	2,332.29	1,562.13	663.01	208.58	82.23	7.50	11.19	37,425.55
15	1,102.91	18,921.17	3,027.46	1,366.27	703.99	766.43	1,198.31	589.17	146.24	69.02	36.00	27,926.98
16	1,004.42	11,251.20	6,634.63	2,117.19	2,194.00	828.36	952.30	354.20	86.39	42.40	6.33	25,471.43
17	1,880.88	4,837.38	6,793.41	765.79	4,638.11	467.33	652.06	244.70	85.79	27.00	9.17	20,401.61
18	6,594.66	466.89	5,819.94	1,126.74	5,886.96	1,391.33	2,109.36	260.71	63.48	36.88	2.48	23,759.42
19	14,641.70	642.09	4,019.68	7,229.49	6,648.35	4,226.77	2,623.11	303.57	52.30	15.35	0.33	40,402.74
20	4,445.11	965.74	1,655.31	2,823.43	2,927.94	2,215.03	2,490.02	692.04	219.39	44.90	2.19	18,481.09
21	4.49	13.33	846.53	3,401.26	3,315.06	2,173.75	1,776.93	1,143.02	337.61	104.04	26.63	13,142.64
	86,542.38	62,559.88	45,342.63	43,180.08	33,944.85	15,509.39	13,406.73	4,220.44	1,222.65	398.47	582.07	306,909.55

Average Shrimp Trawl Fishing Days (24-hour) by Depth Contour (5 fathom intervals) in the Gulf of Mexico, 1998-1999

STATAREA	DEPTH00	DEPTH01	DEPTH02	DEPTH03	DEPTH04	DEPTH05	DEPTH06	DEPTH07	DEPTH08	DEPTH09	DEPTH10	STAT AREA
1	-	16.05	145.68	240.33	7.27	3.03	-	7.71	-	-	7.92	427.99
2	-	13.84	202.65	4,980.86	3,178.94	101.97	4.69	0.00	-	-	37.97	8,520.91
3	-	8.09	95.27	502.08	1,358.25	8.59	1.67	-	-	1.82	-	1,975.76
4	8.09	14.00	580.73	267.11	56.07	-	-	-	-	-	-	926.00
5	11.90	4.96	189.85	991.88	143.34	2.17	-	-	-	-	-	1,344.11
6	-	128.88	1,155.48	1,610.69	59.51	-	-	11.94	-	-	-	2,966.51
7	669.43	580.41	1,031.20	608.27	158.97	15.97	33.31	0.69	-	-	1.51	3,099.76
8	334.48	151.96	346.62	244.65	1,407.07	113.29	129.09	4.22	6.28	-	0.60	2,738.27
9	54.27	0.06	15.90	25.52	198.39	47.00	61.32	2.37	15.52	-	5.67	426.02
10	2,184.46	123.07	214.37	186.97	156.24	83.29	56.60	36.43	11.24	4.42	280.52	3,337.60
11	3,823.20	588.66	1,992.62	4,224.04	767.91	258.56	51.61	24.92	19.06	5.83	1.73	11,758.14
12	6,229.63	390.86	851.30	124.70	41.61	5.71	10.93	8.33	7.79	-	-	7,670.87
13	8,720.23	8,386.85	2,527.14	750.35	507.14	436.83	117.21	168.25	115.34	29.34	6.29	21,764.97
14	9,330.26	8,993.58	2,808.53	1,049.05	644.26	617.66	283.00	114.06	21.16	9.08	0.29	23,870.92
15	420.12	3,834.42	1,152.55	642.25	523.77	290.65	399.34	300.07	163.13	34.35	3.77	7,764.42
16	1,193.32	7,536.04	1,270.91	1,249.66	1,911.23	177.67	359.15	525.22	243.73	32.39	26.87	14,526.17
17	1,444.49	2,052.98	6,798.93	1,099.86	3,296.68	430.25	609.80	227.90	48.03	20.28	16.30	16,045.51
18	5,227.78	204.78	6,679.46	524.61	744.11	265.20	340.80	190.01	74.68	30.57	3.33	14,285.31
19	4,226.81	28.30	2,859.54	3,065.57	6,106.84	2,360.77	2,783.05	732.57	245.56	38.66	12.62	22,460.30
20	1,005.41	81.90	465.32	1,227.13	1,640.36	1,323.07	1,842.90	591.26	231.05	25.28	-	8,433.67
21	-	9.00	224.25	1,001.16	1,816.90	1,044.21	1,431.37	494.27	188.15	23.49	0.50	6,233.30
	44,883.88	33,148.69	31,608.33	24,616.73	24,724.85	7,585.87	8,515.82	3,440.26	1,390.73	255.49	405.88	180,576.51

Percent Change in Average Shrimp Trawl Fishing Days (24-hour) by Depth Contour (5 fathom intervals) in the Gulf of Mexico, 1991-1993 to 1998-1999

STATAREA	DEPTH00	DEPTH01	DEPTH02	DEPTH03	DEPTH04	DEPTH05	DEPTH06	DEPTH07	DEPTH08	DEPTH09	DEPTH10	STAT AREA AVERAGE
1	N/A	416%	3643%	330%	64%	N/A	N/A	20%	N/A	N/A	N/A	481%
2	N/A	N/A	412%	-28%	132%	259%	-97%	-100%	-100%	N/A	11155%	0%
3	N/A	-88%	-84%	-2%	-5%	-21%	N/A	N/A	N/A	N/A	N/A	-25%
4	-61%	-49%	-43%	277%	85%	N/A	N/A	N/A	N/A	N/A	-100%	-37%
5	-43%	5000%	904%	707%	44%	N/A	-100%	N/A	N/A	N/A	-100%	347%
6	-100%	-63%	1481%	310%	52%	-100%	N/A	N/A	N/A	N/A	N/A	227%
7	-15%	15%	354%	-78%	1084%	N/A	1565%	N/A	N/A	N/A	415%	-27%
8	-14%	243%	41%	-75%	2074%	65%	209%	-89%	-79%	-100%	N/A	42%
9	-79%	-90%	86%	37%	892%	726%	2528%	N/A	N/A	N/A	1492%	38%
10	-2%	35%	-63%	-41%	-21%	-19%	-54%	-9%	-7%	342%	1466%	-10%
11	-33%	-4%	-59%	-22%	0%	-56%	-67%	-64%	-11%	N/A	-99%	-36%
12	-51%	27%	232%	-13%	-70%	-92%	-81%	-78%	61%	N/A	-100%	-45%
13	-50%	-52%	-35%	-67%	-55%	-56%	-69%	-27%	46%	4%	779%	-50%
14	-46%	47%	-41%	-76%	-72%	-60%	-57%	-45%	-74%	21%	-97%	-36%
15	-62%	-80%	-62%	-53%	-26%	-62%	-67%	-49%	12%	-50%	-90%	-72%
16	19%	-33%	-81%	-41%	-13%	-79%	-62%	48%	182%	-24%	324%	-43%
17	-23%	-58%	0%	44%	-29%	-8%	-6%	-7%	-44%	-25%	78%	-21%
18	-21%	-56%	15%	-53%	-87%	-81%	-84%	-27%	18%	-17%	34%	-40%
19	-71%	-96%	-29%	-58%	-8%	-44%	6%	141%	370%	152%	3686%	-44%
20	-77%	-92%	-72%	-57%	-44%	-40%	-26%	-15%	5%	-44%	-100%	-54%
21	-100%	-32%	-74%	-71%	-45%	-52%	-19%	-57%	-44%	-77%	-98%	-53%
	-48%	-47%	-30%	-43%	-27%	-51%	-36%	-18%	14%	-36%	-30%	-41%

FIGURE B.10 Shrimp trawl effort (days fished) in the Gulf of Mexico: Comparison of 1991–1993 and 1998–1999 effort by depth and statistical area. These tables provide data on the average number of 24-hour days fished in each of the 21 statistical areas and 10 depth zones in the Gulf of Mexico for 1991–1993 and 1998–1999. For the earlier period an average of 306,909 24-hour fishing days was expended per year. In 1998–1999, this decreased to an average of 180,576 24-hour days per year. Although the greatest portion of the fishing effort in both periods occurred in the shallower depths, 0–20 m, the intensity of trawling per unit area has not always followed the same pattern. The percentage change between the two periods shows a significant decrease in fishing days for most areas and depth zones.

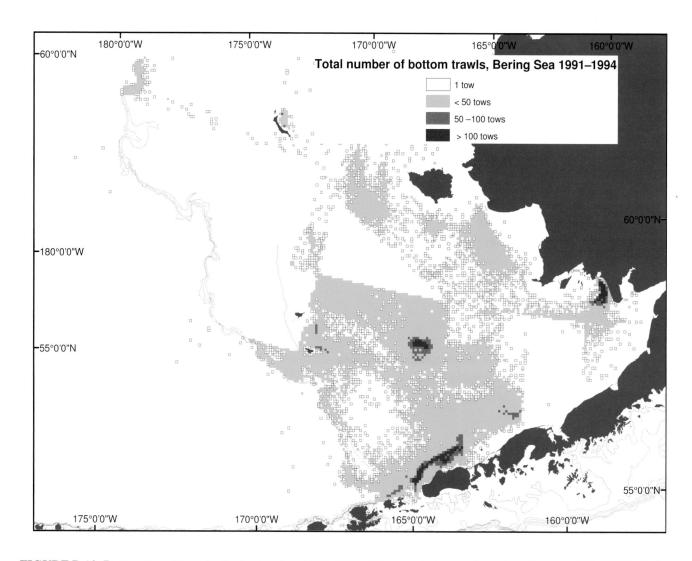

FIGURE B.12 Bottom trawl tows in the Bering Sea, 1991–1994. The cumulative number of trawl tows (1991–1994) within each 25 km² statistical block are mapped 25 km² grid for the Bering Sea. Therefore, the average number of tows per year is about one-fourth the values shown. Even during this period of relatively heavy bottom trawling, tows were not observed for large areas of the Bering Sea.

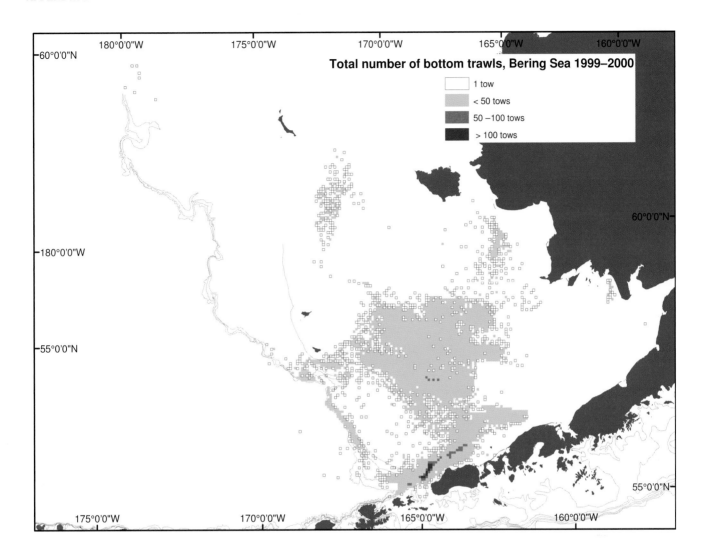

FIGURE B.13 Bottom trawl tows in the Bering Sea, 1999–2000. The effort distribution data for 1999–2000 shows that the total trawl effort declined toward the end of the decade. The average number of observed tows per year in the Bering Sea during the first half of the decade (1991–1994) was 18,925; at the end of the decade (1999–2000) the average was 13,266. This presumes that the sampling effort by vessel classes was similar between periods. There was a reduction in the number of fished cells compared to the first half of the decade and there was a significant reduction in the number of blocks having tows exceeding 100 per year. The reduction reflects North Pacific Fishery Management Council regulations, including those requiring use of midwater trawl gear.

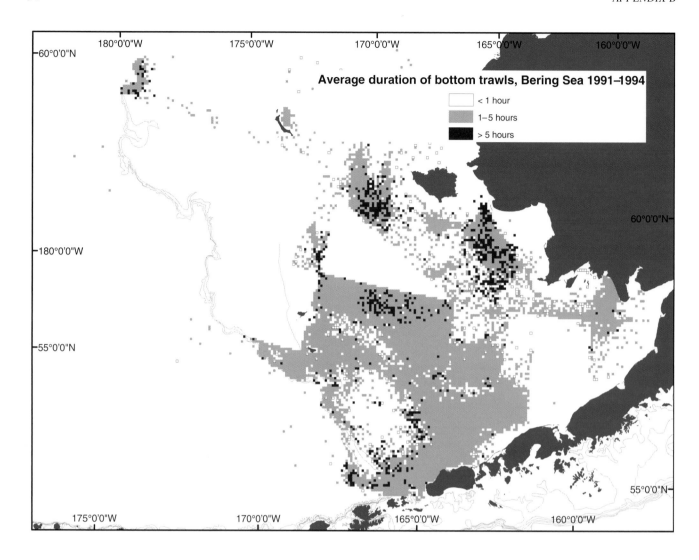

FIGURE B.14 Duration of bottom trawl tows in the Bering Sea, 1991–1994. In the first half of the decade, most of the tows averaged 1–5 hours in length, although for some statistical blocks the average exceeded five hours.

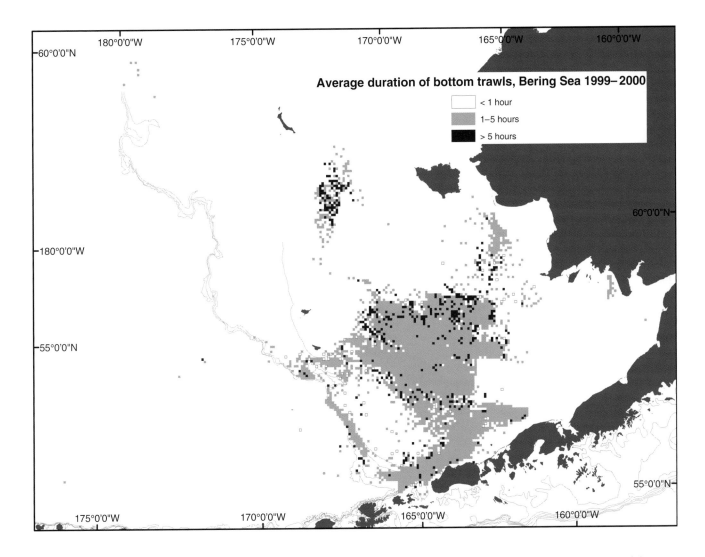

FIGURE B.15 Duration of bottom trawl tows in the Bering Sea, 1999–2000. As in the 1991–1994 period, most of the tows averaged 1–5 hours in length, although for some statistical blocks the average exceeded five hours. There was a dramatic decrease in long duration bottom tows compared to 1991–1994, especially in the northern Bering Sea fishing grounds.

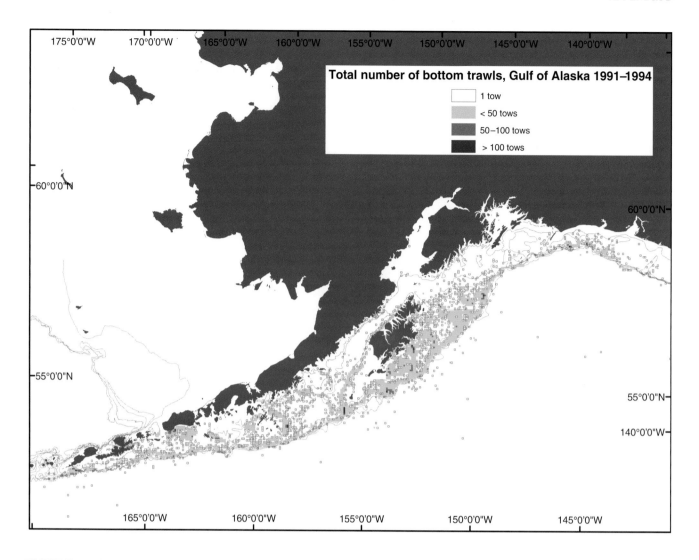

FIGURE B.16 Bottom trawl tows in the Gulf of Alaska, 1991–1994. The Gulf of Alaska does not contain distinct blocks where there were large numbers of hauls relative to the Bering Sea. A few scattered grids show 50–100 tows over the period 1991–1994, but they do not necessarily represent areas where trawling occurred consistently each year.

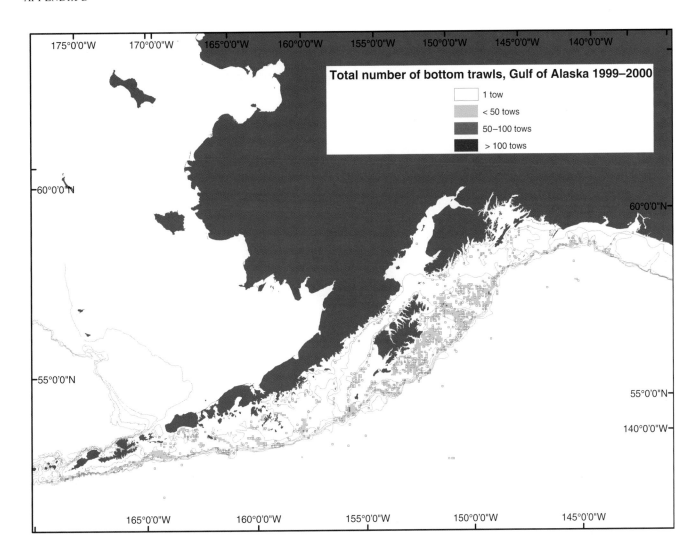

FIGURE B.17 Bottom trawl tows in the Gulf of Alaska, 1999–2000. In the 1999–2000 distribution of tows, the map shows many areas where only one tow occurred, especially within the gullies south of Kodiak Island. In 1999–2000 there were major portions of the shelf and slope that did not have observed tows. The average observed density of trawling in most of the blocks was below 3.25 tows per year (Figure B.24). This reduction in effort, in part, reflects the establishment of no-trawling zones, including the area east of the 140°W longitude line.

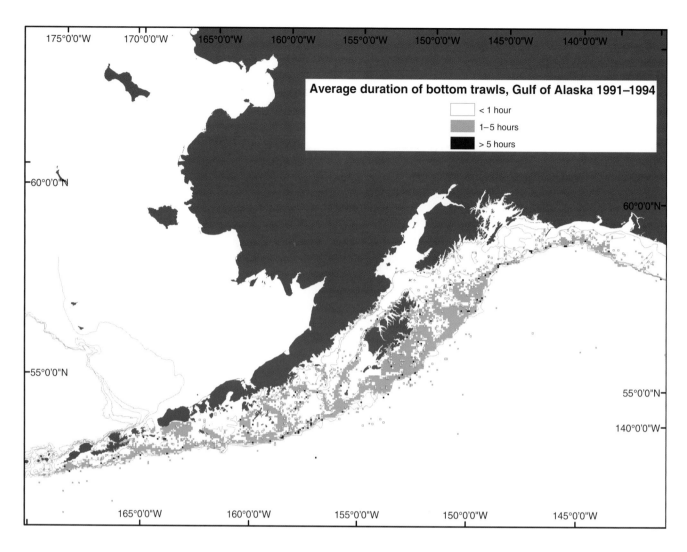

FIGURE B.18 Duration of bottom trawl tows in the Gulf of Alaska, 1991–1994. The length of tows generally ranged 1–5 hours. The average number of observed tows per year were 6842 in 1991–1994. The total (observed plus non-observed) number of tows per year could be significantly higher.

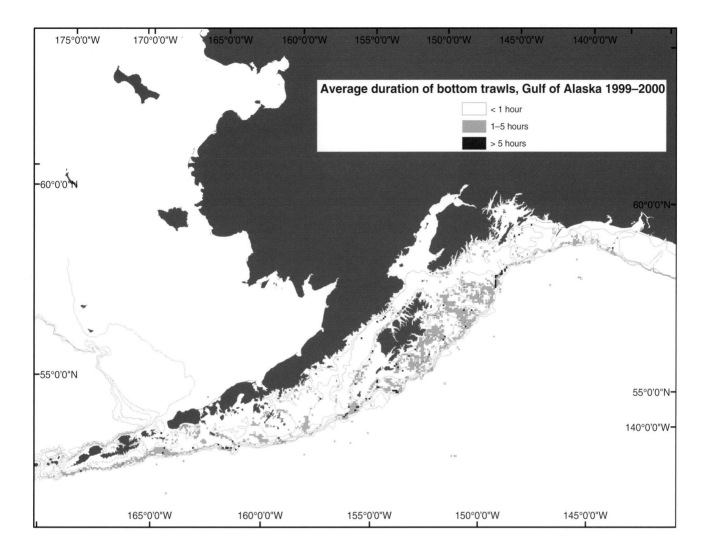

FIGURE B.19 Duration of bottom trawl tows in the Gulf of Alaska, 1999–2000. At the end of the decade, the length of tows generally ranged 1–5 hours. The average number of observed tows per year decreased to 3456 in 1999–2000 compared to 1991–1994. However, the total number of tows per year (observed plus non-observed) could be more than twice these amounts.

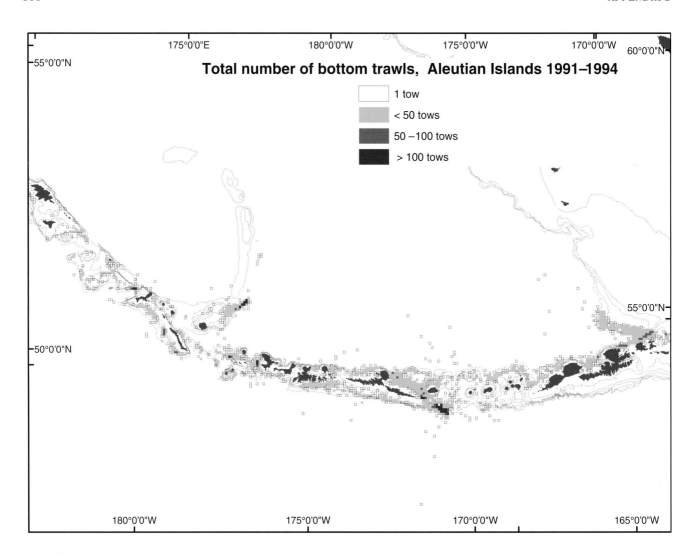

FIGURE B.20 Bottom trawl tows in the Aleutian Islands, 1991–1994. Bottom trawling extends west along the Aleutian Islands from Unimak Island to about 168°E. Trawling intensity is lighter than in the Bering Sea, but distinct concentrations of effort occurred during 1991–1994 in the Sequam Pass area, on Petrel Bank, and off Tanaga Island. There was a relatively large number of single tows for the period, scattered throughout the Aleutian Islands area.

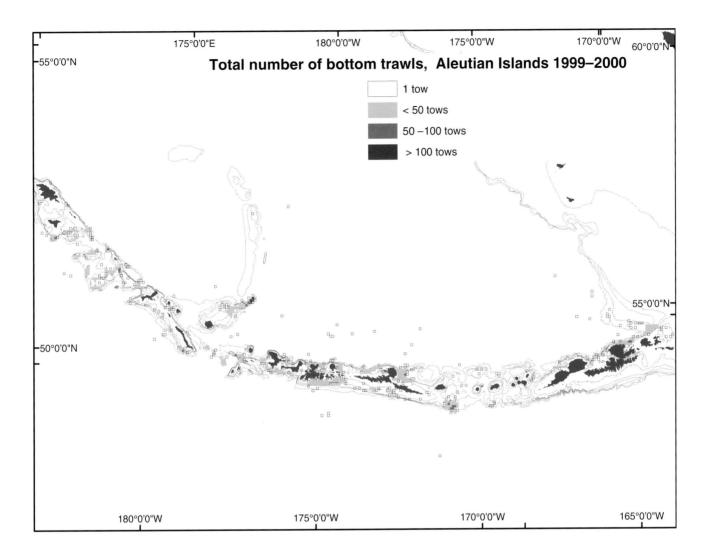

FIGURE B.21 Bottom trawl tows in the Aleutian Islands, 1999–2000. A reduction in bottom trawling effort was observed in 1999–2000. Only a few blocks had more than 100 tows, the number of single-tow blocks declined, and the number of blocks with fewer than 50 tows was greatly reduced. For 1999–2000, no trawling was observed for extensive regions of the shelf and slope. The reduction largely reflects trawl closures intended to protect endangered populations of the Steller sea lion.

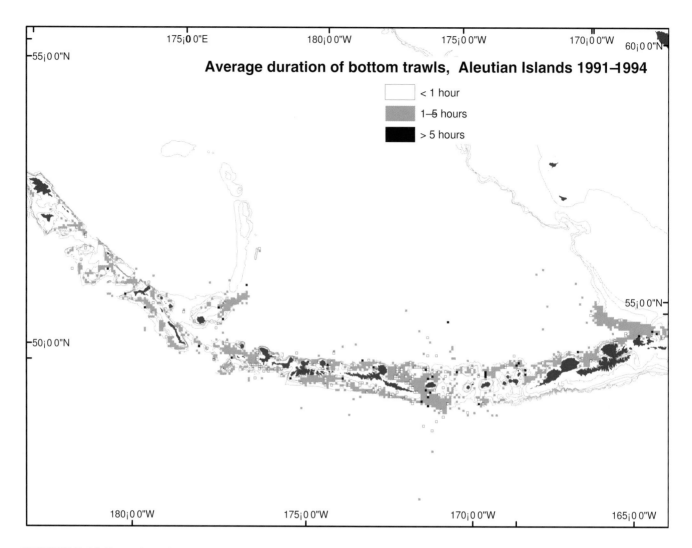

FIGURE B.22 Duration of bottom trawl tows in the Aleutian Islands, 1991–1994. The duration of most tows was typically 0–5 hours. The average number of observed tows per year during 1991–1994 was 4058.

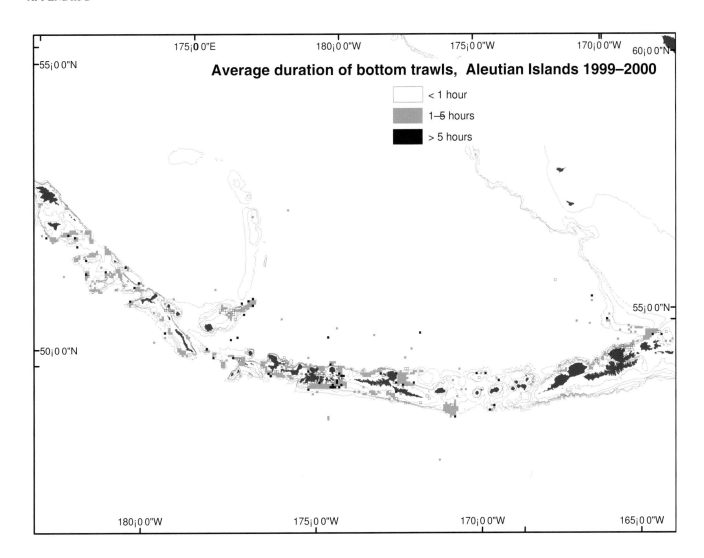

FIGURE B.23 Duration of bottom trawl tows in the Aleutian Islands, 1999–2000. The duration of most tows was typically 0–5 hours. The average number of observed tows per year in 1999–2000 was 2738. It is estimated that only 75 percent of the tows were observed and that the average number of tows during 1999–2000 was 3650.

Bering Sea 1991-1994

	# of grids (25 km^2)	Total # of tows	Approximate area (km^2)	Approximate density (# of tows/25 km^2)
1 tow	1903	1903	47575	1
< 50 tows	5135	46558	128375	9
50 - 100 tows	170	11716	4250	69
> 100 tows	96	15524	2400	162

Bering Sea 1999-2000

	# of grids (25 km^2)	Total # of tows	Approximate area (km^2)	Approximate density (# of tows/25 km^2)
1 tow	1040	1040	26000	1
< 50 tows	2708	21988	67700	8
50 - 100 tows	34	2314	850	68
> 100 tows	9	1190	225	132

Gulf of Alaska 1991-1994

	# of grids (25 km^2)	Total # of tows	Approximate area (km^2)	Approximate density (# of tows/25 km^2)
1 tow	1361	1361	34025	1
< 50 tows	2327	19761	58175	8.5
50 - 100 tows	65	4469	1625	69
> 100 tows	13	1777	325	137

Gulf of Alaska 1999-2000

	# of grids (25 km^2)	Total # of tows	Approximate area (km^2)	Approximate density (# of tows/25 km^2)
1 tow	682	682	17050	1
< 50 tows	863	5655	21575	6.5
50 - 100 tows	8	575	200	72
> 100 tows	0	0	0	0

Aleutian Islands 1991-1994

	# of grids (25 km^2)	Total # of tows	Approximate area (km^2)	Approximate density (# of tows/25 km^2)
1 tow	572	572	14300	1
< 50 tows	831	7727	20775	9
50 - 100 tows	36	2471	900	69
> 100 tows	27	5461	675	202

Aleutian Islands 1999-2000

	# of grids (25 km^2)	Total # of tows	Approximate area (km^2)	Approximate density (# of tows/25 km^2)
1 tow	278	278	6950	1
< 50 tows	413	3647	10325	9
50 - 100 tows	18	1304	450	72
> 100 tows	2	247	50	123.5

FIGURE B.24 Density of trawl fishing off Alaska. This table summarizes the observed bottom trawl tows for the Bering Sea, Gulf of Alaska, and Aleutian Islands for 1991–1994 and 1999–2000. These data were used to approximate areal coverage and the density of tows.

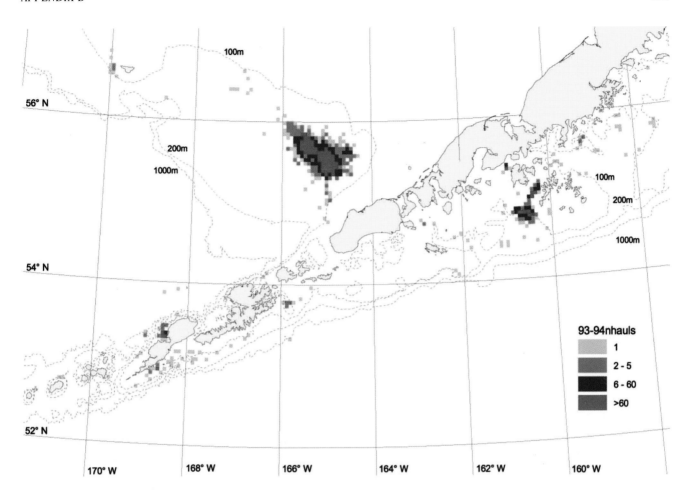

Eastern Aleutians Scallop Dredge Effort
1993-94 Number of Hauls

FIGURE B.25 Scallop dredge effort in the eastern Aleutian Islands, 1993–1994. Most of the scallop dredge effort is concentrated in the Bering Sea north of Unimak Pass (northeast of Dutch Harbor). Fishing depths vary, but generally range from 55 m to 128 m.

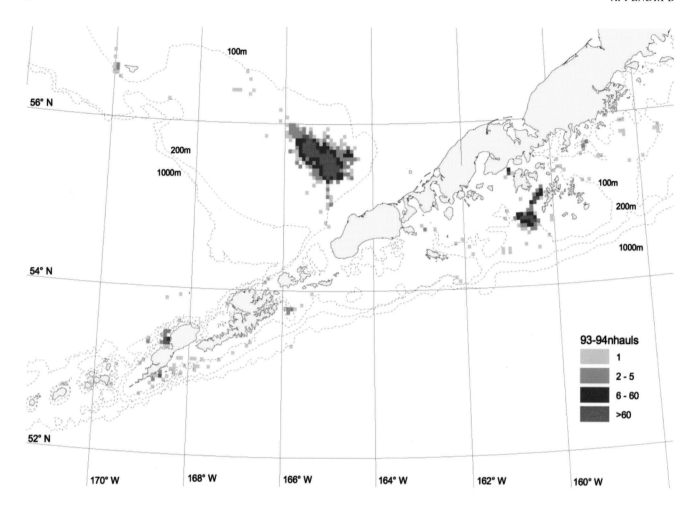

Eastern Aleutians Scallop Dredge Effort
1993-94 Number of Hauls

FIGURE B.26 Scallop dredge effort in the eastern Aleutian Islands, 1999–2000. The density distribution of scallop dredge tows is similar to that seen in 1993–1994, but at lower effort levels.

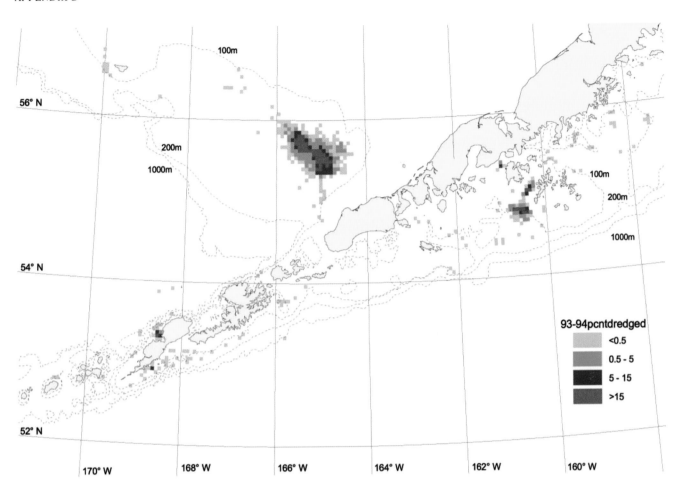

Eastern Aleutians Scallop Dredge Effort
1993-94 Percent Dredged

FIGURE B.27 Area dredged in the eastern Aleutian Islands, 1993–1994. The percentage of area in each 25 km^2 grid box swept by scallop dredge gear parallels the effort data, showing the highest level of coverage in the area north of Unimak Pass. In most blocks, less than 15 percent of the area is swept by dredge gear.

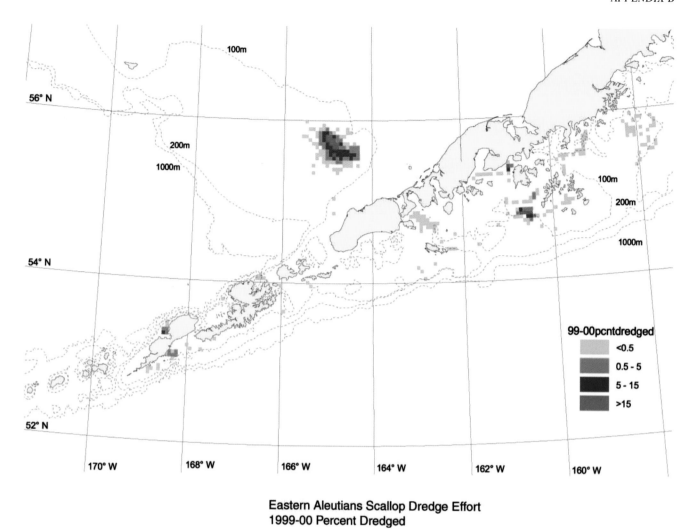

Eastern Aleutians Scallop Dredge Effort
1999-00 Percent Dredged

FIGURE B.28 Area dredged in the eastern Aleutian Islands, 1999–2000. This figure shows the reduction in the amount of area swept by dredge gear in 1999–2000 compared relative to 1993–1994.

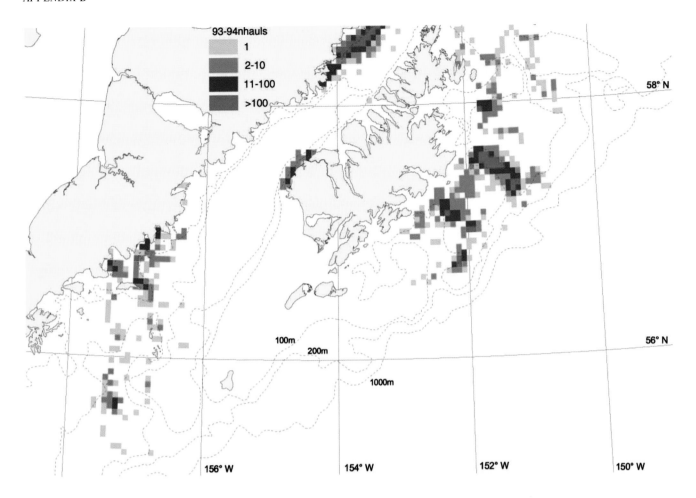

Kodiak Area Scallop Dredge Effort
1993-94 Number of Hauls

FIGURE B. 29 Scallop dredge effort in the Kodiak Area, 1993–1994. Scallop dredge effort is concentrated in the Gulf of Alaska southeast of Kodiak Island and in the Shelikof Straits close to the mainland between Kodiak Island and the mainland. Most dredging occurs at depths <100 m.

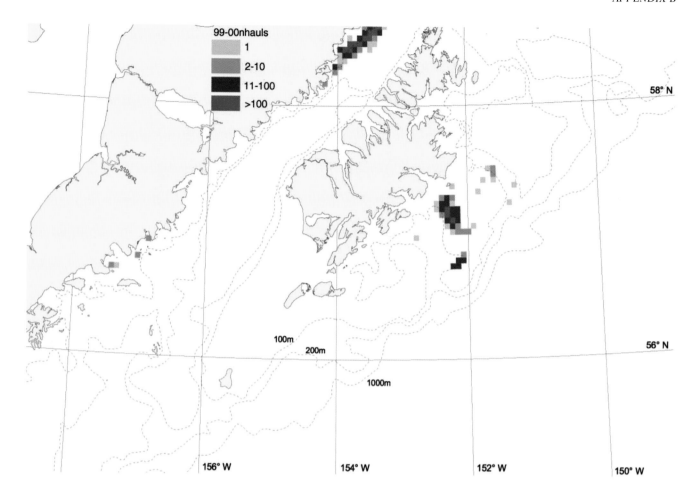

Kodiak Area Scallop Dredge Effort
1999-00 Number of Hauls

FIGURE B.30 Scallop dredge effort in the Kodiak Area, 1999–2000. The density distribution of scallop dredge tows is similar to that seen in 1993–1994, but at lower effort levels.

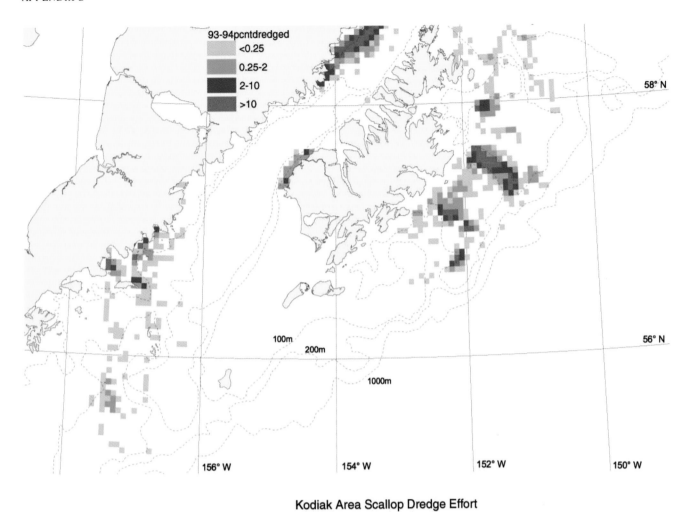

Kodiak Area Scallop Dredge Effort
1993-94 Percent Dredged

FIGURE B.31 Area dredged in the Kodiak Area, 1993–1994. The percentage of area in each 25 km² grid box swept by scallop dredge gear shows a similar distribution as the effort data. Few areas have more that 10 percent coverage by scallop dredging.

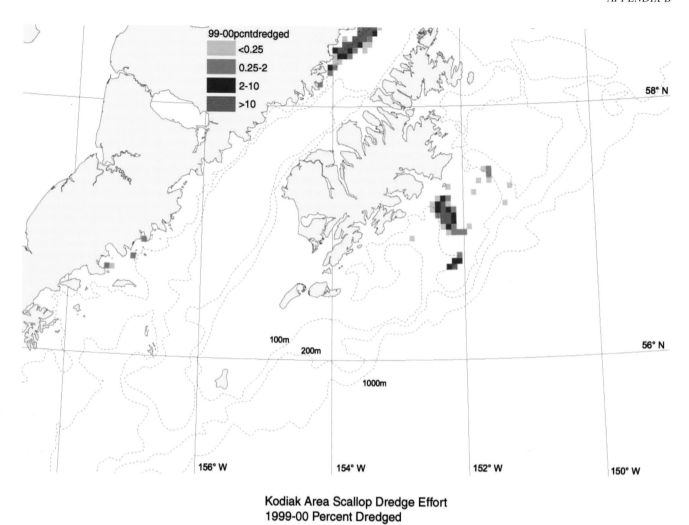

Kodiak Area Scallop Dredge Effort
1999-00 Percent Dredged

FIGURE B.32 Area dredged in the Kodiak Area, 1999–2000. This figure shows the reduction in the amount of area swept by dredge gear in 1999–2000 compared relative to 1993–1994.

CONTIGUOUS WEST COAST STATES (FIGURES B.33–B.37)

Information on the trawling effort off California, Oregon, and Washington was provided by the Natural Resources Consultants, the California Department of Fish and Game, the Oregon Department of Fish and Wildlife, and the Washington Department of Fish and Wildlife. The data were compiled and graphed by Karma Dunlop (Natural Resources Consultants). Oregon and Washington data are 1991–1993 and 1998–1999; California data are 1994–1996. The calculated effort distribution data and the number of tows and duration from all three states are based on log books. Some of the tows are not included in this data set because they could not be geographically resolved to the data cells. Shrimp trawl data are provided for 1979–1999. In August 2001, a compulsory observer program was implemented for California, Oregon, and Washington (Randy Fisher, Pacific State Marine Fisheries Commission, personal communication).

The intensity of trawling off the contiguous West Coast states appears relatively similar for the three states, with perhaps somewhat higher effort occurring off Washington and Oregon (the years for which data have been summarized for Washington, Oregon, and California differ). The average number of tows per year off California during 1994–1996 was 15,535. Although we do not have a 1998–1999 database for California, it is obvious from the Oregon and Washington data that the trawl effort along the West Coast declined sharply between the early 1990s and the latter part of the decade. For example, the number of blocks in the four highest effort categories off Oregon and Washington declined 36 percent, or from 102 to 68 blocks. The decline in effort is demonstrated even more dramatically by the reduction in the average number of tows per year. Between the 1991–1993 and 1998–1999 periods the number of tows declined from 28,489 to 11,487, a reduction of 60 percent in the average number of tows per year. The reduction in effort has followed declines in the abundance of target species and subsequent Pacific Fishery Management Council reductions in quotas and fishing time. The decline continued for all three states in 2000–2001 (Randy Fisher, Pacific State Marine Fisheries Commission, personal communication).

Swept-area information for the region has not been published. To provide an estimate of the affected area, preliminary calculations were performed assuming 3-hour towing times at 3 knots/hour, and a 300-foot door spread. Under these assumptions, one tow would sweep 16.4 million square feet. An average statistical block between 41° and 42° is roughly 71.7 square miles or 2.64 billion square feet. Based on these calculations, it would take about 160 three-hour tows to sweep one statistical block.

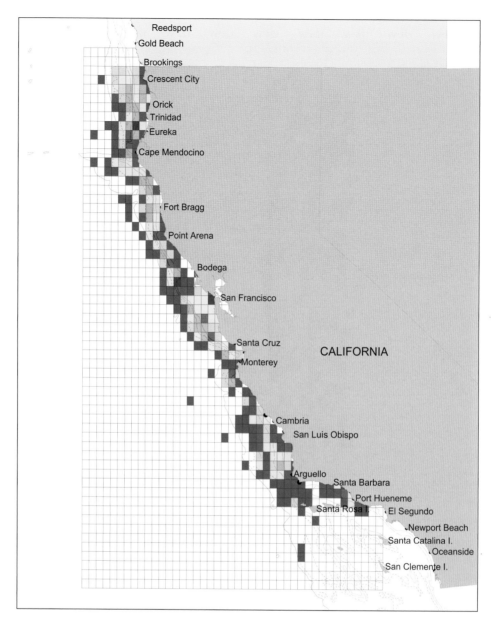

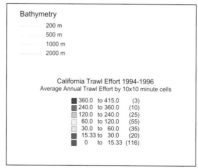

FIGURE B.35 Trawl effort off California, 1994–1996. Important trawl grounds off California, for the most part, are found from Monterey north to the Oregon border, and there is relatively intense fishing between Santa Cruz and San Francisco and between Cape Mendocino and Crescent City. Two relatively heavily fished cells also are reported off Grover City and Ventura. State regulations prohibit trawl fishing on the continental shelf south of El Segundo to the Mexican border, although some trawling occurs on the offshore banks of Southern California. No trawling is allowed within state waters (3 nautical miles).

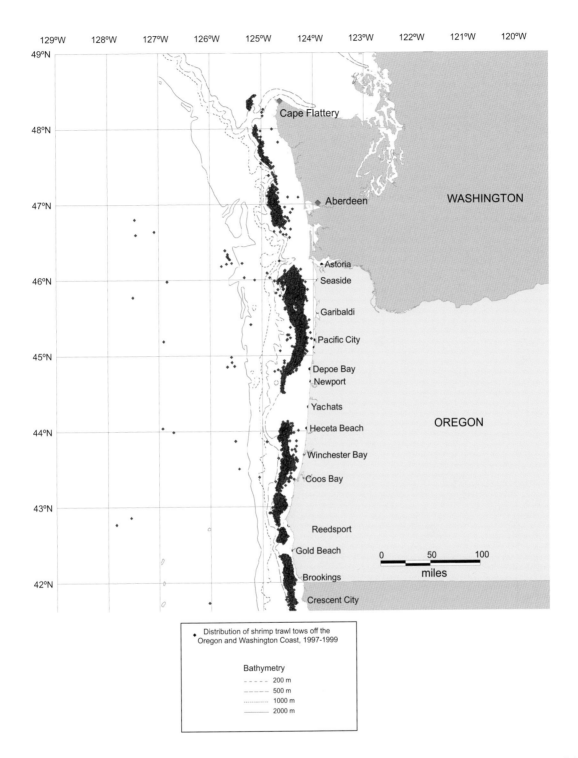

FIGURE B.36 Distribution of shrimp trawls off Oregon and Washington, 1997–1999. There is a significant trawl fishery for northern shrimp. The general distribution of shrimp trawling off Oregon and Washington 1997–1999 is shown. During 1997–1999, an average of 10,000 shrimp tows per year was reported for areas off Oregon and Washington.

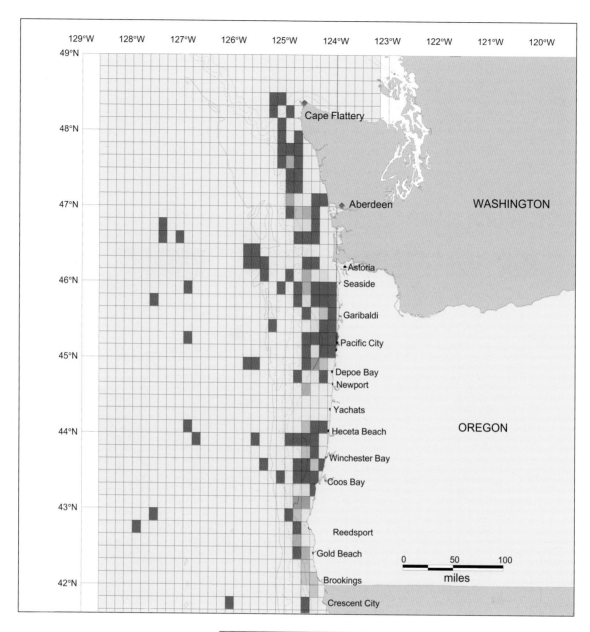

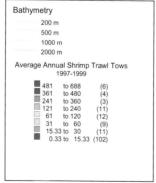

FIGURE B.37 Density of shrimp trawl effort off Oregon and Washington, 1997–1999. This map shows the range of effort in 10′ × 10′ statistical blocks. The most intensive effort occurs off Oregon in the vicinity of Coos Bay, between Seaside and Garibaldi, and off Pacific City.

DECLINE IN TRAWL EFFORT (FIGURE B.38)

Although scientists do not have trawl effort data for all NMFS regions, sufficient data are available to document a considerable decline in trawling in Alaska, Oregon and Washington, and the Gulf of Mexico between the early 1990s (1991–1994) and later in the decade (1998–2000).

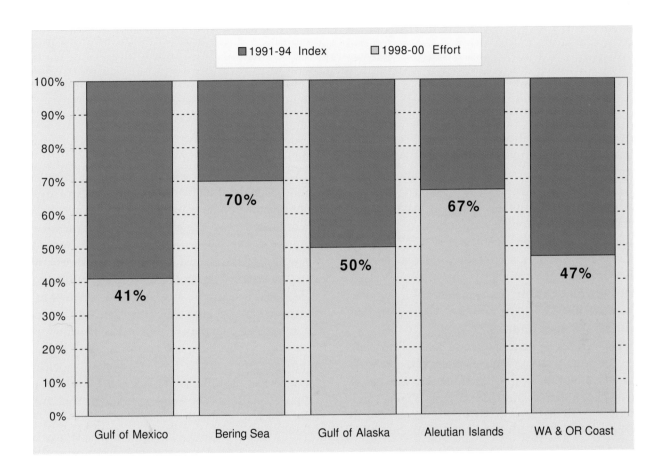

FIGURE B.38 Trawl effort for U.S. trawl fisheries by region, 1991–1994 and 1998–2000. The percentage decline in fishing effort (number of tows or fishing days) were calculated for the Gulf of Mexico, Bering Sea, Gulf of Alaska, Aleutian Islands, and Washington and Oregon. The 1998–2000 effort is depicted relative to the effort level in the 1991–1994 period.

Appendix C

Mapping Tools

Several types of seafloor mapping tools are available, and choosing the right one can be confusing to the nonspecialist. The three types of mapping systems most often applied to seafloor habitats are sidescan-sonar, multibeam, and automatic bottom classification systems. Although a complete discussion of the characteristics, strengths, and weaknesses of each is beyond the scope of this report, a brief discussion is included because it is important to explain that no tool by itself can answer all scientific or management questions. Appropriate mapping tools must be chosen based on the specifics of each case.

Sidescan-sonar systems project a lobe of acoustic energy into the water column that insonifies a wide swath of seafloor on either side of the tow vehicle (or towfish), but a narrow area fore and aft (Figure C.1). Individual scans are accumulated along a trackline to produce a strip of imagery, which resembles a photograph of the seafloor. Adjacent strips often can be assembled into a mosaic to provide a continuous-coverage image of an area. The resolution of the system depends on several factors: the frequency of the signal, the altitude above the seafloor, and the speed and stability of the tow vehicle. A typical system appropriate for shelf mapping might be able to discern objects with dimensions of a few tenths of a centimeter over a range of 100 m (total swath = 200 m). Because of the low incidence angle of the sonar signal, sidescan-sonar systems are particularly good at detecting objects on the seafloor and at showing changes in substrate texture. However, they cannot be used to make bathymetric maps, and because the incidence angle changes across the swath, the interpretation of the data

is somewhat qualitative, making automatic classification of bottom types difficult.

Multibeam echosounders use numerous sonar beams at a high angle of incidence with the seafloor and produce narrow swaths of bathymetric data, which can be accumulated into complete-coverage bathymetric maps (Figure C.1). Some systems also can record the strength of the reflected signal ("backscatter"), which can give information about the hardness or texture of the substrate. Systems are available for a range of resolutions and water depths. The consistent high-incidence beam creates quantitative depth data that are conducive to data processing and bottom classification. The high-incidence beam, however, makes multibeam systems less useful than are sidescan systems for discerning small objects, such as boulders that are less than 1 m across.

Automatic bottom classification systems (e.g., RoxAnn and QTC-View) are single-beam echosounders that analyze the character of the return echo from a footprint directly under the ship. Before being heard, they must be calibrated in a well-known area that has bottom types similar to those in the survey area. Data from separate surveys can be difficult to compare. Because they do not insonify a swath of seafloor, creating a map from the data requires considerable interpolation, and significant features can be missed. They are relatively simple for a nonspecialist to use.

The character of the substrate and the questions being addressed will dictate the choice of mapping tool. In an area where the seafloor is relatively smooth and where distinguishing gravel bed from fine-grained sediments is important, a relatively low-resolution

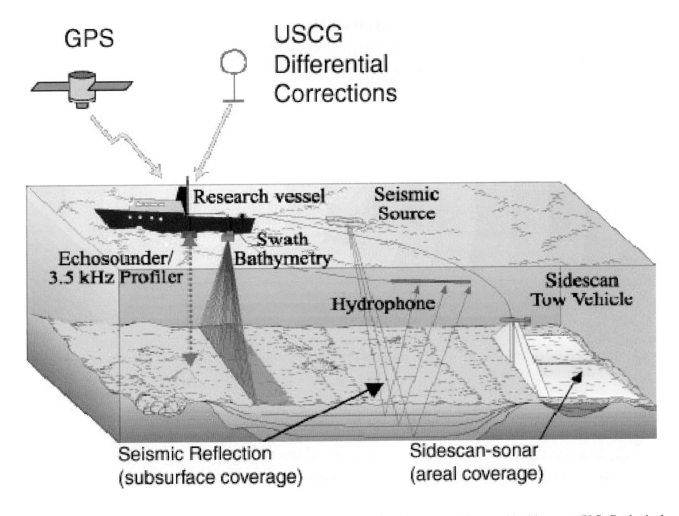

FIGURE C.1 Typical configurations of various types of seafloor mapping instruments discussed in this report (U.S. Geological Survey, 2001).

multibeam bathymetry system with backscatter data recorded might be ideal. In an area where individual boulders are important habitats, a sidescan system would probably be a better choice. If the seafloor is of uniform composition, but subtle changes in depth are important, a high-resolution multibeam system is appropriate. An automatic bottom classification system might be used as a preliminary reconnaissance tool to expand from a well-known area into an unknown area. The use of supporting data—from substrate samples or photographic or video observations—is necessary to interpret data from any of these mapping tools in terms of habitat.

Appendix D

Workshop Agendas

New England Aquarium
Central Wharf
Boston, MA 02110
February 5-7, 2001

MONDAY, FEBRUARY 5

OPEN SESSION

11:00 a.m.	**Welcome**—John Steele, *Chair*, and Susan Roberts, *Study Director*
11:15 p.m.	**Sponsor perspectives**—Michael Sissenwine, *National Marine Fisheries Service*
12:00 p.m.	**Lunch**
1:00 p.m.	Jeremy Collie presents overview of effects of mobile gear on seafloor habitats
1:40p.m.	Peter Auster presents overview of effects on benthic communities
2:20 p.m.	Joe DeAlteris presents overview of gear versus natural disturbance
3:00 p.m.	**Open session adjourns for the day**

TUESDAY, FEBRUARY 6

OPEN SESSION

8:00 a.m.	**Breakfast**
8:30 a.m.	**Welcome and introduction**—John Steele, *Chair*
9:00 a.m.	**Gear Effects**
	Overview—Joe DeAlteris
9:15 a.m.	**Types of bottom fishing gear and interaction with the seabed**—Arnie Carr, *Massachusetts Department of Fisheries*
9: 45 a.m.	**Sediment resuspension by mobile fishing gear**—Jim Churchill, *Woods Hole Oceanographic Institution*
10:15 a.m.	**Break**

10:30 a.m.	**The effect of bottom tending mobile gear in New England: A conservation perspective**—Anthony Chatwin, *Conservation Law Foundation*
11:00 a.m.	**Problems with the environmental assessments of bottom trawling and scallop dredging**—Chris Zeman, *American Oceans Campaign*
11:30 a.m.	**General Discussion**
12:00 p.m.	**Lunch**— **Preliminary balance sheet for Georges Bank closed areas**—Brian Rothschild, *University of Massachusetts, Dartmouth*
1:00 p.m.	**Gear effect and habitat studies in Georges Bank closed areas**—Frank Almeida, *Northeast Fisheries Science Center, NMFS*
1:30 p.m.	**Habitat concerns of the New England Fishery Management Council**—Tom Hill, *Chair, New England Fishery Management Council*
2:00 p.m.	**Marine habitat management in Maine**—Linda Mercer, *Maine Department of Marine Resources*
2:30 p.m.	**Management considerations for implementing habitat protection measures**—Andy Rosenberg, *University of New Hampshire*
3:00 p.m.	**Break**
3:15 p.m.	**A fisherman's perspective on trawling**—Bill Amaru, *Fisherman and New England Fishery Management Council Member*
3:45 p.m.	**Fishing gear and impacts of small boat trawlers**—David Goethel, *Fisherman*
4:00 p.m.	**Trawling and dredging effects: An industry perspective**—Nils Stolpe, *Garden State Seafood Association*
4:30 p.m.	**General Discussion**
5:30 p.m.	**Meeting adjourns for the day**

Moody Gardens Hotel and Conference Center
7 Hope Boulevard
Galveston, TX 77554
March 5-6, 2001

MONDAY, MARCH 5

OPEN SESSION

1:00 p.m.	**Welcome and introduction**—John Steele, *Chair*

Gulf of Mexico Fisheries

1:30 p.m.	**Management of bottom gear in the Gulf of Mexico with emphasis on trawling and thoughts on impacts to essential fish habitat**—Rick Leard, *Gulf of Mexico Fishery Management Council*
2:00 p.m.	**State management of bottom trawl fisheries**—Jeff Rester, *Gulf States Marine Fisheries Commission*
2:30 p.m.	**Discussion**
3:00 p.m.	**Break**

Community Concerns

3:30 p.m.	**Recommendations for addressing the impacts of bottom trawling in the Gulf of Mexico**—Chris Dorsett, *Gulf Restoration Network*

4:00 p.m.	**Recommendations to reduce the negative effects of trawling on Gulf of Mexico essential fish habitat**—Pam Baker, *Environmental Defense*
4:30 p.m.	**The effects of bottom trawling on seafloor habitats of the western Gulf of Mexico: Trawling refugia and ecosystem considerations**—Benny Gallaway, *Texas Shrimp Association*
5:00 p.m.	**General discussion**
5:30 p.m.	**Meeting adjourns for the day**
5:30 p.m.	**Reception in the Garden Terrace for committee members, speakers, and attendees**

TUESDAY, MARCH 6

OPEN SESSION

8:00 a.m.	**Breakfast**
8:30 a.m.	**Welcome and introduction**—John Steele, *Chair*

Fishing Gear and Habitat

9:00 a.m.	**Shrimp trawl impacts on the bottom**—David Harrington, *University of Georgia*
9:30 a.m.	**Distribution of shrimp trawling effort in the Gulf of Mexico**—Pete Sheridan, *NMFS Galveston*
10:00 a.m.	**Break**
10:15 a.m.	**Developing research priorities for the Southeast Region**—Alonzo Hamilton, *NMFS Pascagoula*
10:45 a.m.	**Review of fishing gear used in the Southeast Region and their potential impact on essential fish habitat**—Mike Barnette, *NMFS St. Petersburg*
11:15 a.m.	**Oculina Banks**—Felicia Coleman, *Florida State University*
11:30 a.m.	**General discussion**
12:00 p.m.	**Lunch**
1:00 p.m.	**Open session adjourns for the day**

WestCoast International Inn
Anchorage, AK
May 31-June 2, 2001

FRIDAY, JUNE 1

OPEN SESSION

12:00 p.m.	**Lunch**
12:30 p.m.	**Overview of North Pacific ecosystem and fisheries**—Gordon Kruse, *Alaska Department of Fish and Game*
1:00 p.m.	**Management perspectives on effects of bottom trawling on marine habitat off Alaska**—Chris Oliver, Dave Witherell, and Cathy Coon, *North Pacific Fishery Management Council*
1:30 p.m.	**A trawl industry perspective on the seafloor effects of trawling off Alaska**—John Gauvin, *Groundfish Forum*; Bill Hayes, *Commercial Fisherman*
2:10 p.m.	**Commercial flatfish fishing in the Bering Sea: Experimental trawling and implications for long-term change**—Eloise Brown, *University of Alaska, Fairbanks*
2:30 p.m.	**Trawl dynamics and its potential impact on habitat**—Gary Loverich, *NET-systems*
2:50 p.m.	**Break**